Intelligent Systems Reference Library

Volume 181

The aim of this series is to publish a Reference Library, including novel advances and developments in all aspects of Intelligent Systems in an easily accessible and well structured form. The series includes reference works, handbooks, compendia, textbooks, well-structured monographs, dictionaries, and encyclopedias. It contains well integrated knowledge and current information in the field of Intelligent Systems. The series covers the theory, applications, and design methods of Intelligent Systems. Virtually all disciplines such as engineering, computer science, avionics, business, e-commerce, environment, healthcare, physics and life science are included. The list of topics spans all the areas of modern intelligent systems such as: Ambient intelligence, Computational intelligence, Social intelligence, Computational neuroscience, Artificial life, Virtual society, Cognitive systems, DNA and immunity-based systems, e-Learning and teaching, Human-centred computing and Machine ethics, Intelligent control, Intelligent data analysis, Knowledge-based paradigms, Knowledge management, Intelligent agents, Intelligent decision making, Intelligent network security, Interactive entertainment, Learning paradigms, Recommender systems, Robotics and Mechatronics including human-machine teaming, Self-organizing and adaptive systems, Soft computing including Neural systems, Fuzzy systems, Evolutionary computing and the Fusion of these paradigms, Perception and Vision, Web intelligence and Multimedia.

** Indexing: The books of this series are submitted to ISI Web of Science, SCOPUS, DBLP and Springerlink.

More information about this series at http://www.springer.com/series/8578

Christos Troussas · Maria Virvou

Advances in Social Networking-based Learning

Machine Learning-based User Modelling and Sentiment Analysis

Springer

Christos Troussas
Department of Informatics
University of Piraeus
Piraeus, Greece

Maria Virvou
Department of Informatics
University of Piraeus
Piraeus, Greece

ISSN 1868-4394 ISSN 1868-4408 (electronic)
Intelligent Systems Reference Library
ISBN 978-3-030-39129-4 ISBN 978-3-030-39130-0 (eBook)
https://doi.org/10.1007/978-3-030-39130-0

This Springer imprint is published by the registered company Springer Nature Switzerland AG
The registered company address is: Gewerbestrasse 11, 6330 Cham, Switzerland

Foreword

Even though *artificial intelligence in education* is a research field of long and ever increasing attention, there seems to be deficiencies in the development of personalized software to meet resultant needs and challenges. On the other hand, while it has arisen as an evolution of previous ways of e-learning, *social networking-based learning* remains to date a somewhat unexplored research area. Christos Troussas and Maria Virvou have investigated these important and hot research areas that have attracted the interest of the scientific community, namely artificial intelligence in education and social networking. The incentive of their research lies in the fact that the incorporation of artificial intelligence techniques in social networking-based learning can enhance the development of individualized and adaptive software which is tailored to the cognitive needs and personal preferences of the students in a social learning environment. Indeed, these technologies have made significant advances recently and have become burning issues with increasing research projects in academia and IT industry.

This book at hand reflects the important effort of the authors to address these emerging challenges and constitutes a noteworthy addition to the related field. The authors employ an intriguing approach that incorporates machine learning, conceptual frameworks, pedagogical and learning theories, natural language processing, decision-making methodologies and sentiment analysis techniques in order to further promote individualization and adaptivity in social e-learning environments and refine the learner–computer interaction. Moreover, the authors develop a broad paradigm built using the aforementioned models, which is analyzed in each chapter of this book.

In view of the above, I consider this work a significant improvement for both the areas of artificial intelligence in education and learning through social networking. I am confident that this book will help researchers and software developers to create better educational software, offering greater personalization and adaptivity to the stakeholders and ameliorating, in parallel, human–computer interaction in educational environments.

Prof.-Dr. George A. Tsihrintzis
University of Piraeus
Piraeus, Greece

Preface

This book covers three very important and hot research issues, namely the social networking-based learning, the machine learning-based user modeling and the sentiment analysis. These three technologies have been widely used worldwide by researchers and serve for the design of academic disciplines along with the development of R&D departments of the IT industry. However, these three technologies have not been used extensively yet for the purposes of education. Thus, the authors of this book present a novel approach that uses adaptive hypermedia in e-learning models to personalize educational content and learning resources based on the needs and preferences of individual learners. According to reports of 2019, the vast majority of internet users worldwide are active social networking users and the global average social network penetration rate as of 2019 is close to half the population. The gains of utilizing the technology of social networking in the field of education can reap the advantages of new technological advancements in order to create interactive educational environments where students can learn, collaborate with peers, communicate with tutors while benefiting from a social and pedagogical structure similar to a real class which does not impose physical presence. In this way, using social networks for education is a very powerful and promising idea.

This book first presents in depth the current trend of social networking-based learning, as an evolution of previous ways of e-learning. Then, it provides a novel framework that moves further from digital learning technologies while incorporating a wide range of current advances to provide solutions to future challenges. The described novel approach incorporates machine learning to the student modeling component which also uses conceptual frameworks and pedagogical theories in order to further promote individualization and adaptivity in e-learning environments. Moreover, error diagnosis to misconceptions, tailored testing and collaboration between students are examined in depth and novel approaches for these modules are presented. Sentiment analysis is also incorporated in the general framework supporting personalized learning, by considering the user's emotional state, and creating a user-friendly learning environment tailored to students' needs. Reinforcement to students, in the form of motivation, completes the framework and assists students in their educational effort. As such, this book will help researchers of knowledge-based software engineering to build more sophisticated personalized educational software, while retaining a high level of adaptivity and user-friendliness within human–computer interaction. Furthermore, educators and software developers can be given significant information when designing and implementing intelligent tutoring systems and adaptive educational hypermedia systems.

Piraeus, Greece

Dr. Christos Troussas
Dr. Maria Virvou

Contents

1 Introduction 1
1.1 Current Topics 1
1.2 Social Networks as Learning Tools 4
1.2.1 Facebook 4
1.2.2 Google+ 6
1.2.3 Twitter 7
1.2.4 Elgg 8
1.2.5 Edmodo 9
1.3 Comparative Analysis 10
1.3.1 Criteria 11
1.3.2 Results for Using SNs in Educational Contexts 11
1.4 Related Fields and Open Research Questions 14
References 15

2 Related Work 17
2.1 Social Media Language Learning 17
2.2 Related Literature for Social e-Learning 19
2.2.1 Methodology and Model Used 19
2.2.2 Selected Systems in the Review 20
2.2.3 Multicriteria Framework for Social e-Learning Systems 22
2.3 Comparative Discussion 27
References 29

3 Intelligent, Adaptive and Social e-Learning in POLYGLOT 33
3.1 Intelligent Tutoring Systems (ITSs) 33
3.1.1 Architecture of ITS 35
3.1.2 Function of ITSs 36
3.2 Intelligent Computer-Assisted Language Learning (ICALL) 37
3.3 User Modeling and Adaptivity 40

3.3.1 Student Models Characteristics 41
3.3.2 Using a Student Modeling in an ITS 45
3.4 Platforms for Social e-Learning 45
3.4.1 Research Method 48
3.4.2 Comparative Analysis and Discussion 50
References 55

4 Computer-Supported Collaborative Learning: A Novel Framework 59
4.1 Computer-Supported Collaborative Learning (CSCL): An Introduction 59
4.2 Precursor Theories 60
4.3 Collaboration Theory and Group Cognition 61
4.4 Strategies 61
4.5 Instructor Roles in CSCL 63
4.6 Effects 63
4.7 Applications of CSCL 64
4.8 CSCL for Foreign Language Acquisition 65
4.8.1 Effectiveness and Perception 65
4.9 Win-Win Collaboration Module 66
References 67

5 Affective Computing and Motivation in Educational Contexts: Data Pre-processing and Ensemble Learning 69
5.1 Affective Computing 69
5.1.1 Affective States 70
5.2 Frustration as an Affective State 71
5.2.1 Rosenweig's Frustration Theory 72
5.2.2 Frustration Aggression Hypothesis 72
5.2.3 Frustration and Goal-Blockage 73
5.2.4 Frustration and Cause in Computer Users 73
5.3 Motivation Theory 74
5.3.1 Hull's Drive Theory 74
5.3.2 Lewin's Field Theory 75
5.3.3 Atkinson's Theory of Achievement Motivation 75
5.3.4 Rotter's Social Learning Theory 76
5.3.5 Attribution Theory 76
5.3.6 Discussion on Motivational Theories 77
5.4 Responding to Frustration 78
5.5 Pre-processing Techniques and Ensemble Classifiers for Sentiment Analysis Through Social Networks 79
5.5.1 Methodology 79
5.5.2 Twitter Datasets 81

5.5.3 Evaluation of Data Preprocessing Techniques 81
5.5.4 Evaluation of Stand-Alone Classifiers 88
5.5.5 Evaluation of Ensemble Classifiers 91
References 95

6 Blending Machine Learning with Krashen's Theory and Felder-Silverman Model for Student Modeling 99
6.1 Employing the Stephen Krashen's Theory of Second Language Acquisition in POLYGLOT 99
6.2 POLYGLOT Learning Content 101
6.3 POLYGLOT Student Model 104
6.3.1 Approximate String Matching for Error Diagnosis 107
6.3.2 String Meaning Similarity for Error Diagnosis 108
6.4 Automatic Detection of Learning Styles Based on Felder-Silverman Model Using the K-NN Algorithm 109
6.5 Tailored Assessments 111
6.5.1 Criteria for Tailored Assessment 113
6.5.2 Overview of the Building of the Adaptive Test Algorithm 115
References 118

7 Regression-Based Affect Recognition and Handling Using the Attribution Theory 121
7.1 Declaration and Handling of Affective States 121
7.2 Automatic Detection of Frustration 123
7.3 Rectilinear Regression Model to Detect Frustration 124
7.4 Incorporation of the Rectilinear Regression Model in POLYGLOT 125
7.5 Respond to Frustration 127
7.6 Delivery of Motivational Messages Based on the Attribution Theory 128
References 134

8 Overview of POLYGLOT Architecture and Implementation 135
8.1 POLYGLOT Architecture 135
8.2 POLYGLOT Implementation 138
References 150

9 Evaluation Results for POLYGLOT and Discussion 153
9.1 Evaluation Process and Framework Used 153
9.1.1 Criteria 156
9.1.2 Method 156
9.1.3 Population 158

9.2 Results 159
9.2.1 Satisfaction 159
9.2.2 Performance 159
9.2.3 Individual State of Learners 163
9.2.4 Students' Progress 165
9.2.5 Validity of the Detection of the Students' Learning Style 165
9.2.6 Validity of Win-Win Collaboration 166
9.3 Questionnaires 166
References 168

10 Conclusions 169
10.1 Conclusions and Discussion 169
10.2 Contribution to Science 173
10.2.1 Contribution to Intelligent Tutoring Systems 173
10.2.2 Contribution to Computer-Supported Collaborative Learning 174
10.2.3 Contribution to Student Modeling 174
10.2.4 Contribution to Computer-Assisted Language Learning 175
10.2.5 Contribution to Affective Computing 175
10.3 Future Work 176
References 176

Chapter 1
Introduction

Abstract This chapter includes introductory aspects of e-learning and intelligent tutoring systems and generally the application of artificial intelligence in the field of education. More specifically, it introduces topics pertaining to the employment of cognitive and learning theories, learning style models, machine learning algorithms, statistical modeling and decision analysis towards creating a personalized and adaptive learning experience for the students. The aforementioned techniques are incorporated in a novel social networking-based language learning system, called POLYGLOT that we have fully developed and evaluated. Moreover, the chapter provides an insight on how social networks can serve as learning tools and presents open research questions on the related fields, proving that there is scope for a lot of improvement in terms of adaptivity and personalization in social networking-based learning.

1.1 Current Topics

The world has witnessed major improvements in the areas of Information Technology and telecommunications. These important changes have permitted the rise of the phenomenon of globalization by which regional economies, societies, and cultures have become integrated through a global network of people. As a result, all the emerging needs of modern life accentuate the importance of digital learning (e.g. online learning of foreign languages) [1, 2]. Considering the scientific area of Intelligent Tutoring Systems (ITSs), there is an increasing interest in the use of computer-assisted foreign language instruction [3]. In this way, students may learn a foreign language, by using a computer-assisted application. Especially, when these systems offer the possibility of multiple-language learning at the same time, the students may further benefit from this educational process [4].

In recent years, the rapid development of high and new technology has opened new horizons in computer-assisted instruction. Intelligent Tutoring Systems are based on computer models of instructional content and support the learning, by providing personalized instruction to students. In this way, students may learn one or more foreign languages. European reality necessitates multiple language learning, so the

C. Troussas and M. Virvou, *Advances in Social Networking-based Learning*, Intelligent Systems Reference Library 181, https://doi.org/10.1007/978-3-030-39130-0_1

students may further benefit from this educational process. For this reason, the need of systems that incorporate intelligence is even greater when students are taught more than one foreign language simultaneously [5].

One important area of ITSs involves the specialization on language learning which is referred to as Intelligent Computer-Assisted Language Learning (ICALL). In ICALL, students are taught a language (e.g. Greek, English, French etc.) through an ITS. Nowadays, all the emerging needs of everyday life along with the phenomenon of globalization accentuate the significance of learning foreign languages. Moreover, it has to be emphasized that foreign language learning is widely promoted by many countries and clusters of countries. For example, the European Union promotes such guidance for its country members. Due to the currents global promotion of language learning, countries, such as Greece, have adopted foreign language teaching in the education curriculum of schools. Students are obliged to learn two foreign languages starting from the primary school to the secondary school. The teaching of foreign languages (English, French and German) is compulsory for all European pupils in all three grades. Even though the English and French languages have common characteristics so that their learning can be joint [6, 7], there is the risk of students being confused in multiple language learning.

The need for tutoring systems that may provide user interface friendliness and also individualized support to errors via a student model are even greater when students are taught more than one foreign languages simultaneously [3]. A solution to this problem may be the integration of the technology of Intelligent Tutoring Systems (ITSs), so as to provide adaptive tutoring to individual students. ITSs offer intelligence and adaptivity to individual students' needs, via student modeling. The individual student model for each student contains information about the knowledge level and the error handling of the student in each concept of multiple language learning. Hence, error diagnosis is a module which supports the students while studying theory and solving exercises [8]. Socialization has important pedagogical implications in collaborative learning that support the learners' personal relationships and social interaction with their classmates [9, 10]. Therefore, the support of collaboration in multiple language learning may promote the learning process. When adaptive personalized e-learning systems could accelerate the learning process by revealing the strengths and weaknesses of each student in a collaborative environment, they could dynamically plan lessons and personalize the communication and didactic strategy [11]. Machine learning techniques can be used for acquiring models of individual users interacting with educational systems and group them into communities or stereotypes with common interests [12], so that the student reap the benefits of collaboration.

Collaboration has helped humans realize shared goals, especially in cases where individual effort has been found inadequate. Over the last years we have all witnessed the power of groups working together and the electronic human networks that are changing the way we see the World Wide Web [13]. Correspondingly, collaboration is quite recently used in electronic learning software to help people involved in a common e-learning task achieve goals [14]. It is believed that humans as social beings have an endogenous tendency to create groups. Many scientists in the area of educational learning support that it would be educationally highly beneficial if these groups could consist of learners that would work complementarily [15, 16].

Social networking sites (SNSs) (e.g. Facebook, Twitter, MySpace) have become commonplace interactivity tools that bring people together through computer-based approaches. The main features of SNSs that render them very popular over other means of online communication include immediacy, interactivity, and self-identification development through continuous engagement with one another [17]. Studies have showed that social network tools support educational activities by enhancing interaction, collaboration, active participation, information and resource sharing, and critical thinking [18–21]. Current research on social networks (SNs) has focused on identity, network structures, privacy and technological issues; therefore, there is the recognizable need for research on social networks in educational contexts [18, 22]. However, research on social networking in intelligent educational contexts is still limited.

Social networks seem particularly useful for the purposes of language learning through computer-assisted education. Troussas et al. [23] point out that socialization has important pedagogical implications in language learning that support the learners' personal relationships and social interaction with their classmates.

SNSs can make the learning of a second language through socialization faster. The social networks offer people the facility to be surrounded by the target language, to have sufficient interaction and to actively participate in discussions [17, 24]. On the other hand, a very crucial element in language learning is the learner centeredness, pedagogical approach and learner's autonomy. Inevitably, learners must be at the center of teaching pedagogical practices [25], taking into consideration their needs, preferences, interests or even their sentiments [26, 27]. Learners' autonomy has been attributed to many definitions, such as the ability to take charge of one's own learning, a capacity—for detachment, critical reflection, decision-making and independent action [28], and recognition of the rights of learners within educational systems [17].

In view of the above, the main goal of this research is to profit from the features of social networks and the technology of ITSs by combining them in a novel way in order to offer optimized and personalized multilingual learning. Given that European students are obliged to learn two foreign languages since primary school due to European regulation, the teaching of foreign languages (English and French) is integrated in the curriculum. English is compulsory for all pupils in all three grades, while pupils can choose French, as a second compulsory option.[1] Towards this direction, as a testbed for our research, this work focuses on developing a prototype system for learning grammatical phenomena in English and French, as foreign languages. The system, named POLYGLOT, is a web-based intelligent tutoring system with social characteristics, such as posting on a wall, tagging a classmate, instant and asynchronous text messaging, declaration of the affective state, reaction buttons in exercises, student group collaboration. Furthermore, it involves the generation of personalized recommendation for collaboration, which is adapted to users' needs, the diagnosis of users' quiz misconceptions, the automatic detection of students'

[1] http://www.greeceindex.com/greece-education/greek_education_foreign_languages.html.

learning style assisting them in their learning experience and the automatic detection of students' frustration and a response on it in order to ameliorate the tutoring process. In particular, POLYGLOT incorporates the following:

- the Stephen Krashen's Theory of Second Language Acquisition, that involves features, such as the way of instruction, means of collaboration, time constraints in learning, holding students' records, logical gradation of learning concepts and response on negative affective state (frustration) in the form of motivational messages
- the Felder-Silverman Learning Style Model, for determining the students' learning styles
- a supervised machine learning algorithm (k-nearest neighbors algorithm) which takes as input several students' features, including their age, gender, educational level, computer knowledge level, number of languages spoken and grade on preliminary test, in order to detect their learning style
- Approximate String matching for diagnosing types of students' errors
- String meaning similarity for diagnosing errors due to language transfer interference
- techniques for tailored assessment
- a dynamic model for adaptive domain knowledge delivery and personalized assessment units using multiple-criteria decision analysis
- the Linear Regression model to automatically detect students' frustration
- the Attribution Theory to deliver appropriate motivational messages to students.

1.2 Social Networks as Learning Tools

In this section, we describe the SNs selected for review and the way they could be used as learning tools. Furthermore, we display screenshots of the developed learning process through these tools.

1.2.1 Facebook

Facebook[2] is a well-liked social networking site allowing registered users to create personal profiles and connect with other users by sharing their personal opinion in the form of a status, personal photos and other items of interest. Furthermore, it allows users to interact with their peers through synchronous and asynchronous messaging, posting and utilization of reactions buttons, such as "Like", "Sad", etc. Moreover, it involves public features, such as the following:

- Pages; users can create and promote a public page that is related to a specific scope.

[2]https://www.facebook.com/.

- Presence technology; it allows users to see the friends who are online and to chat synchronously.
- Events; users can advertize an event, invite people and track people who are likely to participate.
- Groups; users sharing common interests can communicate with each other.

Educational use

Groups in Facebook can be seen as a valuable tool for assisting the tutoring of any curriculum subject (Fig. 1.1). Files can be uploaded either by the administrator

Fig. 1.1 Learning process through Facebook

of the group or by group members depending on predefined privacy settings. As such, instructors can deliver learning materials and students can communicate with peers, collaborate and learn through chatting, posting and commenting. Furthermore, instructors can create debates on modern and ongoing issues and inform learners about an upcoming exam or an essay deadline, setting up an event on Facebook. Facebook groups can be also utilized by instructors to communicate with parents, providing prompt information on students' progress besides face-to-face meetings.

1.2.2 Google+

Google+ [3] is an internet-based social network that is owned and operated by Google; it was shut down for business use and consumers on April 2, 2019. Google+ continued to be available as "Google+ for G Suite", later rebranded as "Google Currents".[4] A user profile in Google+ is a public visible account of a user including fundamental characteristics of social networking services, such as a profile/background/cover photo, an "about" section showing personal information about the user, such as previous work, education, interests, etc. and an area to post status updates. In addition, it has a "+1 button" allowing people to recommend sites and parts of external sites, similar to the use of Facebook's "Like" button. The core features of this platform are:

- Communities, allowing users to create ongoing conversations about particular topics.
- Circles, enabling users to organize people into groups or lists for sharing ideas.

Educational use

Google+ provides better student collaboration through Circles, which are opportunities for blended learning (a combination of offline and online instruction) using Hangouts, project research using Sparks, and easier school public relations with content sharing and messaging (Fig. 1.2). Instructors can use Google+ to communicate directly with learners, learners' families or other educators. Google+ is integrated with others services, such as Google Calendar and Google Docs, supporting project-based learning (a dynamic classroom approach in which students acquire a deeper knowledge through active exploration of real-world challenges and problems).

---

[3] https://plus.google.com/.

[4] https://gsuite.google.com/products/currents/.

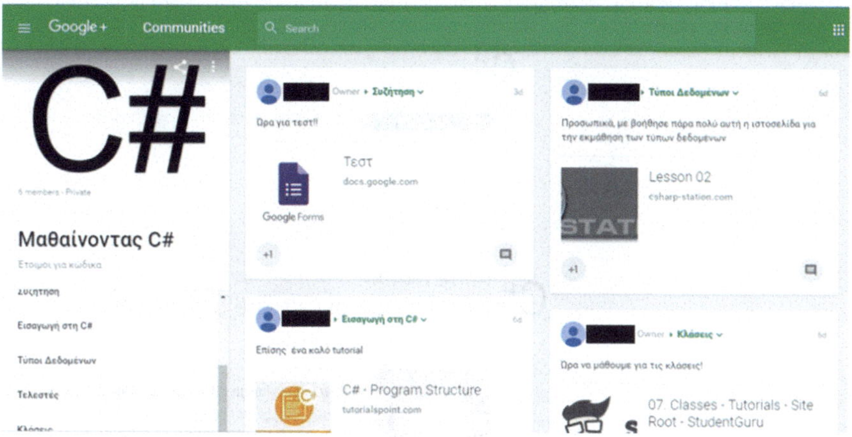

Fig. 1.2 Learning process through Google+

1.2.3 *Twitter*

Twitter[5] is one of the most popular social micro-blogging platforms. As a social network, users post messages interacting with other people; these messages are called "tweets" and are restricted to 140 characters. Users can subscribe to accounts of other users receiving their tweets; this is known as "following", while subscribers are known as "followers". Tweets are publicly visible by default; however, senders have the option to restrict message visibility solely to their followers. Moreover, individual tweets can be forwarded by other users to their own feed, a process known as a "retweet", or they can be "liked" (formerly "favorite"); these two features that provide users' interaction.

Educational use

Twitter proves to be exceptionally useful for supporting e-learning applications. Teachers, students, and parents can greatly benefit from the advantages offered by using Twitter as an educational tool. The short tweets can be used to inform students about any changes pertaining to the curriculum, post a question or request resources about the domain to be taught, share interesting links for a deeper insight of the lesson, and work collaboratively. Custom course hashtags can be set up around lessons and topics, so that students follow these specific hashtags to keep a record of what has been taught during a course (Fig. 1.3).

[5]https://twitter.com/.

Fig. 1.3 Learning process through Twitter

1.2.4 Elgg

Elgg[6] is an open source social networking platform providing a robust framework to develop any kind of social networking sites. Most of the functionalities come from the installation of the proper plugins. It offers blogging, microblogging, file sharing, networking, groups and a number of other features. Each user has its own weblog, file repository (with podcasting capabilities), online profile and RSS (Rich Site Summary) reader.

Educational use

Elgg offers a set of features being appropriate for developing online learning networks, such as weblogs, bookmarks, instant messaging, storing files, sharing

[6]https://elgg.org/.

Fig. 1.4 Learning process through Elgg

resources and connecting with others (Fig. 1.4). Learners have their online profiles where they can maintain their file gallery, blogs, and personal customized templates. Besides that, the Wiki add-on component renders Elgg more practical for an academic course community allowing students to work together. Moreover, this platform can be integrated with other learning management systems, such as Moodle, to produce software oriented to the field of education.

1.2.5 Edmodo

Edmodo[7] is a full-featured social learning platform designed for participants in the educational process wanting to connect and collaborate with their peers. Using this

[7] https://www.edmodo.com/.

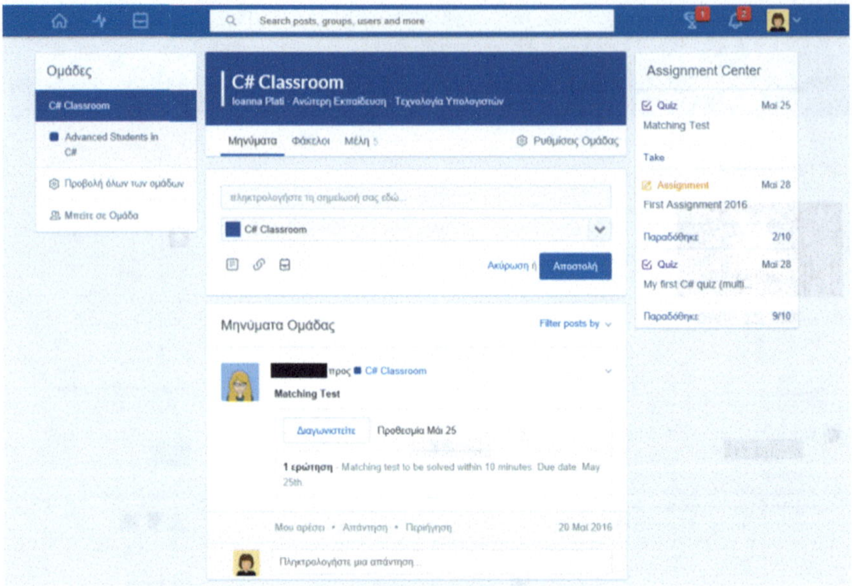

Fig. 1.5 Learning process through Edmodo

platform, instructors and learners as well as their parents or guardians, can connect by sharing opinions, concerns, and helpful tips, sending notes, replying to posts and checking messages, grades and upcoming events, while they are away from the classroom. Edmodo cannot be seen as a learning management system (LMS), but as an instructors-oriented social learning platform.

Educational use

Edmodo focuses on the development of educational SNs. Teachers can create groups, assign homework, schedule quizzes, manage progress, etc. Edmodo gives students new ways to engage, participate, and express themselves. Through posting on discussion topics, participating in polls, being awarded with badges, and more, it is fostered the communication and learning becomes social (Fig. 1.5).

1.3 Comparative Analysis

The scope of this research is to evaluate the use of several well-known SNs as learning tools, providing formal and informal tutoring, and to compare the selected platforms regarding their educational features.

Firstly, we conducted a thorough investigation of the research area, focusing on the characteristics of SNs and their exploitation in the teaching and learning process. Subsequently, as a testbed for this research, we developed a learning process using several well-known SNs, selected for review. The course that has been chosen for teaching was an undergraduate course in the Department of Informatics of the University of Piraeus, namely that of the programming language C# ("Object-Oriented Programming"), which constitutes a lesson in Computer Science Departments of many Universities. Finally, we evaluated the SNs from an educational perspective and analyzed the results.

1.3.1 Criteria

An effective e-learning environment supports a number of features in order to carry out the tasks of learning process. Based on these features, we analyze the capability of SNs to act as learning tools. Thus, the criteria of evaluation are:

- Course management: The capability to create a course and manage its settings.
- Content management: The capability to deliver and manage course material.
- Student management: The capability to manage students and monitor their activities.
- Course Enrollment: The process of initiating attendance to the course.
- User-generated content: The capability of students to create and deliver content.
- Assessment tool: The capability to create online quizzes and assignments, and manage them and their deliverables.
- Gradebook: The capability to store student's grades from all assessment (quiz, exam, essays etc.) and provide reports.
- Communication & Collaboration: The tools that enable the participants of the learning process to communicate and collaborate in this context.
- User-friendliness: if the platform provides an easy-to-use interface.

1.3.2 Results for Using SNs in Educational Contexts

In this study, we conduct a comparative analysis of five well-known SNs in terms of provided educational features. Table 1.1 illustrates the evaluation results.

The emergence of SNs broadens learning's horizons, extending online learning environments to social ones. The exploitation of SNs in education promotes the learning through social interaction and collaboration. Moreover, the daily use of them by learners and instructors in their personal life renders their adoption in learning process more easy.

Facebook is one of the most used SNs and the majority of learners and instructors have a Facebook account for peer communication. This fact fosters its adoption, as there is no need to create a new account or use a different platform for learning, besides its use in everyday life. Furthermore, the users are familiar with its interface making it easier to be used the learning process. Facebook enables a great interaction through its groups.

Table 1.1 Evaluation of SNS

Characteristics	SNs				
	Facebook	Google+	Twitter	Elgg	Edmodo
Course management	Through Facebook groups	Through Google+ communities	Through a Twitter account	Through groups	Only a teacher account can create course pages
Content management	Upload files, images, videos etc. and configure post settings. Manage material using hashtags	Upload files, images, videos etc. and configure post settings. Manage material using tags	Post text, links and images. Manage material using hashtags	Bookmarks section, Uploaded files section, Groups providing group activity/files/blog/discussions	Upload files, images, videos etc. and configure post settings. Organize material in folders
Student management	No	No	No	No	Assign student into groups. No activity reports
Course Enrollment	Join Facebook group	Join community	Follow course/teacher account	Join group	Join group using code
User-generated content	Post material and comment posts	Post material and comment posts	Through tweets	Posts in blog and discussions	Post material, comment posts, Backpack module (a place where student upload their material)
Assessment tool	No	No	No	No	Yes
Gradebook	No	No	No	No	Yes
Communication and Collaboration	Posts, comments, private chatting	Posts, comments, private chatting	Tweets, comments, private chatting	Posts, comments, private messages	Posts, comments, send message to teacher
User-friendliness	Yes	Yes	Yes	Yes	Yes

Twitter is the most representative SN providing microblogging service, with an abundance of users. It has an easy-to-use interface; however, its significant restrictions are the short length of tweets and the lack of uploading files in tweets. Hence, course content should provide as image or link.

Google+ is a well-known SN with less popularity in comparison to the aforementioned ones.[8] However, it provides a great content organization with tags and a user-friendly environment. In addition, it can incorporate other Google services, such as forms, calendar etc.

Elgg is a social networking platform which provides an integrated solution for developing social e-learning environments. It has an easy-to-use interface and enables group creation with special features, for instance blog, files, pages, discussions etc. Using groups, the course content can be delivered effectively. The main disadvantage of this platform is the limit variety of plugins.

Edmodo is social networking platform for developing educational environments. It provides a Facebook-like interface, making it user-friendly. Through this platform, the instructors can develop integrated online learning environments and manage course content and students in an easy way.

In general terms, all the aforementioned SNs, being used for this review, are free of charge and provide a user-friendly environment either for teachers to organize their lessons or students to participate in learning activities. Moreover, they foster social interaction and collaborative learning, promoting innovative forms of teaching and learning.

In contrast to Elgg and Edmodo, Facebook, Google+ and Twitter have an additional advantage which is that the majority of educational community already uses these platforms. Hence, they are familiar with them and it does not need to create new account or get used with different platform.

All the evaluated platforms, except for Edmodo, do not provide education-oriented features such as student tracking, assessments, gradebook etc.; thus they have limited capabilities as learning tools. However, all the platforms facilitate the instructional process, providing a learning experience beyond the traditional classroom.

In view of the above, social networks have the potential to facilitate the interaction between peers. Their use in higher education can enhance the learning experiences of students, promoting co-creation of content, communication and collaboration. This study evaluates five well-known SNs, namely Facebook, Twitter, Google+ , Elgg and Edmodo, as tools for teaching and learning. The main advantage of their adoption in the learning process is that they are free of charge and provide a friendly user interface to which the majority of people is accustomed. Hence they offer easiness in use for both learners and tutors. Furthermore, they support formal and informal, as well as collaborative learning. However, concerning Facebook, Twitter and Google+ , their main drawback is that these platforms lack education-oriented features, thus offering a poor and limited functional learning environment.

---

[8] https://dustn.tv/social-media-statistics/.

1.4 Related Fields and Open Research Questions

This study aims at answering several research questions emerging from the proliferation of technological advancements in the field of web-based instruction tailored to SNSs. All the questions follow the direction of placing the student in the center of the educational process. Hence, the research questions emerging from this study are the following:

1. Can computer science itself assist effectively on digital learning (e.g. learning a foreign language) through the use of social networking characteristics, in a way that learning autonomy is adopted?

   This question is critical because it seeks to investigate if the social features can promote the education and how they can be incorporated to benefit the students. Given that social networks have invaded the everyday life, people, and especially the younger generation, tend to devote a lot of time to communicating through posting on digital walls, sending private messages to peers, commenting and expressing their feeling using corresponding reaction buttons. Thus, this study will give insight on how the aforementioned characteristics can enhance the instruction process.
2. How can the student learning style be predicted automatically using as less characteristics as possible in order to save student's time?

   Defining the learning style model is a cumbersome process and requires answering a lot of questions from student. Hence, the student should invest much time for this purpose. In order to exceed this time restraint, the current work tries to find relationships between student characteristics and learning styles for classifying students according to their style in an algorithmic way. To this direction, it is important to specify the appropriate student characteristics, such as age, gender, educational level etc., and the proper learning style model that will identify the different way with which a student learns.
3. Based on which approach can the system recommend collaborations between users in order to provide effective learning through adequate groups?

   Collaboration between students is an essential module of e-learning systems that can be further promoted through the adoption of social networking features. Towards an efficient collaboration where both students can reap its benefits, the proper approach for collaboration should be identified. As such, the system will be able to recommend those peers from the learning community to students that meet the requirements for a complementary collaboration.
4. How can the error diagnosis mechanism further enhance the tutoring process?

   Error diagnosis, especially in tutoring systems for learning language, is the cornerstone of the education process because there are many different misconception categories concerning grammatical concepts. Firstly, it is necessary to define these error categories and associate them with a variety of explanations about the possible cause of the mistake. After that, the way of elaborating them should be identified for a more individualized instruction.

5. In which way does the students' characterization of the exercises affect the content adaptivity to them?

   The liking or disliking of the exercises by the students serves as an important input to the frustration detection mechanism and can promote a student-centered tutoring process.

6. Does adaptive learning material delivery and personalized assessment to students upgrade students' knowledge level?

   Dynamic course material and personalized assessment can be valuable tools for supporting a student-centered learning environment, taking into consideration the students' needs, preferences and interests.

7. How can the detection of frustration and the response to frustration in the form of motivation assist the learning process?

   The automatic detection of frustration and the response to frustration in the form of motivational messages are very important given that the frustration constitutes an impediment of the learning process that may impel student to quit learning.

References

1. Kurata, N.: Opportunities for foreign language learning and use within a learner's informal social networks. Mind, Cult., Act. **17**, 382–396 (2010)
2. Troussas, C., Espinosa, K.J., Virvou, M.: Intelligent advice generator for personalized language learning through social networking sites. In: 6th International Conference on Information, Intelligence, Systems and Applications (IISA), pp. 1–5. Corfu (2015a)
3. Troussas, C., Virvou, M., Vougiouklidou, A., Espinosa, K.J.: Automatic misconception diagnosis in multiple language learning over social networks. In: International Conference on Information, Intelligence, Systems & Applications (IISA), pp. 1–6. Piraeus (2013a)
4. Troussas, C., Virvou, M., Espinosa, K.J.: Using visualization algorithms for discovering patterns in groups of users for tutoring multiple languages through social networking. J. Netw. **10**(12), 668–674 (2015b)
5. Troussas, C., Virvou, M., Caro, J., Espinosa, K.J.: Mining relationships among user clusters in Facebook for language learning. In: International Conference on Computer, Information and Telecommunication Systems (CITS), pp. 1–5. Athens (2013b)
6. Roberts, I.G.: Verbs and Diachronic Syntax: A Comparative History of English and French. Studies in Natural Language and Linguistic Theory, pp. 1–340. Springer (1993)
7. Vinay, J.P., Darbelnet, J.: Comparative Stylistics of French and English: A Methodology for Translation. Benjamins Translation Library, pp. 1–358 (1995)
8. Virvou, M., Alepis, E., Troussas, C.: Error diagnosis in computer-supported collaborative multiple language learning using user classification. In: Frontiers in Artificial Intelligence and Applications Knowledge-Based Software Engineering—Proceedings of the Tenth Conference on Knowledge-Based Software Engineering, vol. 240, pp. 266–275 (2012)
9. Caballé, S., Xhafa, F., Barolli, L.: Using mobile devices to support online collaborative learning. J. Mob. Inf. Syst. **6**(1), 27–47 (2010)
10. Troussas, C., Virvou, M., Alepis, E.: Collaborative learning: group interaction in an intelligent mobile-assisted multiple language learning system. Inform. Educ. **13**(2), 279–292 (2014)

11. Licchelli, O., Basile, T., Di Mauro, N., Esposito, F., Semeraro, G., Ferilli, S.: Machine learning approaches for inducing student models. Lecture Notes in Artificial Intelligence (Subseries of Lecture Notes in Computer Science). In: 17th International Conference on Industrial and Engineering Applications of Artificial Intelligence and Expert Systems, pp. 935–944 (2004)
12. Papatheodorou, C.: Machine learning in user modeling. In: Paliouras G., Karkaletsis V., Spyropoulos C.D. (eds) Machine Learning and Its Application, Lecture Notes in Computer Science, vol. 2049, pp. 296–294. Springer, Berlin (2001)
13. Benevenuto, F., Rodrigues, T., Cha, M., Almeida, V.: Characterizing user navigation and interactions in online social networks. Inf. Sci. **195**, 1–24 (2012)
14. Lichtnow, D., Gasparini, I., Bouzeghoub, A., De Oliveira, J.P.M., Pimenta, M.S.: Recommendation of learning material through students collaboration and user modeling in an adaptive e-learning environment, dies in computational intelligence. In: Daradoumis, T., Caballé, S., Juan, A.A., Xhafa, F. (eds.) Technology-Enhanced Systems and Tools for Collaborative Learning Scaffolding. Studies in Computational Intelligence, vol. 350, pp. 257–278. Springer, Berlin, Heidelberg (2011)
15. Dafoulas, G.A., Swigger, K., Brazile, R., Alpaslan, F.N., Serce, F.C., Cabrera, V.L.: Supporting collaboration in global software development through e-Learning. In: Proceedings of the IADIS International Conference e-Learning, pp. 201–204 (2009)
16. Troussas, C., Krouska, A., Virvou, M.: Automatic predictions using LDA for learning through social networking services. In: IEEE 29th International Conference on Tools with Artificial Intelligence (ICTAI), pp. 747–751. Boston, MA (2017)
17. Benson, P.: Teaching and Researching Autonomy in Language Learning, pp. 1–296. Longman/Pearson Education, Harlow (2001)
18. Mazman, S.G., Usluel, Y.K.: Modeling educational usage of Facebook. Comput. Educ. **55**(2), 444–453 (2010)
19. Ajjan, H., Hartshorne, R.: Investigating faculty decisions to adopt Web 2.0 technologies: theory and empirical tests. Internet High. Educ. **11**(2), 71–80 (2008)
20. Selwyn, N.: Web 2.0 applications as alternative environments for informal learning—a critical review. Paper presented at the OECD-KERIS expert meeting. Alternative learning environments in practice: using ICT to change impact and outcomes (2007)
21. Mason, R.: Learning technologies for adult continuing education. Stud. Continuing Educ. **28**(2), 121–133 (2006)
22. Lockyer, L., Patterson, J.: Integrating social networking technologies in education: a case study of a formal learning environment. In: Proceedings of 8th IEEE International Conference on Advanced Learning Technologies, pp. 529–533 (2008)
23. Troussas, C., Virvou, M., Caro, J., Espinosa, K.J.: Language learning assisted by group profiling in social networks. Int. J. Emerg. Technol. Learn. (iJET) **8**(3), 35–38 (2013)
24. Krouska, A., Troussas, C., Virvou, M.: Social annotation tools in digital learning: a literature review. In: 9th International Conference on Information, Intelligence, Systems and Applications (IISA), pp. 1–4. Zakynthos, Greece (2018)
25. Li, K.H., Lou, S.J., Tseng, K.H., Huang, H.C.: A preliminary study on the Facebook-based learning platform integrated with blended learning model and flip learning for online and classroom learning. In: Wang, J.F., Lau, R. (eds.) Advances in Web-Based Learning—ICWL 2013. ICWL 2013. Lecture Notes in Computer Science, vol. 8167, pp. 172–183. Springer, Berlin, Heidelberg (2013)
26. Krouska, A., Troussas, C., Virvou, M.: Comparative evaluation of algorithms for sentiment analysis over social networking services. J. Univers. Comput. Sci. **23**(8), 755–768 (2017)
27. Troussas, C., Krouska, A., Virvou, M.: A multicriteria framework for assessing sentiment analysis in social and digital learning: software review. In: 9th International Conference on Information, Intelligence, Systems and Applications (IISA), pp. 1–7. Zakynthos, Greece (2018)
28. Holec, H.: Autonomy and Foreign Language Learning, pp. 1–65. Council of Europe, Oxford, Pergamon and Strasbourg (1981)

Chapter 2
Related Work

Abstract In this chapter, we present the related work in the scientific area of social networking-based learning and we investigate the evolution of intelligent systems in educational contexts as to what extent social networking has affected this area. Social networking-based learning involves either already existing social networking platforms being used as educational tools or intelligent e-learning systems incorporating social characteristics. However, e-learning systems that were implemented using well-known social networking platforms (e.g. Facebook, Twitter, Elgg etc.) tend to be based mainly on their social character and they do not offer yet student-centered instruction in terms of adaptive and individualized learning. Moreover, these systems do not incorporate yet expert decision making, namely decisions on teaching strategies, diagnosis and inference mechanisms. In contrast, intelligent e-learning systems tend to mainly incorporate intelligence in their reasoning and diagnostic mechanisms but they offer very limited social characteristics in the educational process. This chapter provides an insight to the aforementioned research area that can be valuable to tutors, senior and junior researchers and software developers in the field of digital learning. Also, this chapter can serve as a guideline for incorporating intelligent and expert techniques promoting personalization and adaptivity in the field of e-learning and specifically in social networking-based learning.

2.1 Social Media Language Learning

Social Media Language Learning (SMLL) links to language learning through interactive social media channels. This allows students to develop language skills and communication skills. Social media consists of interactive media forms that enable users to interact with each other and publish, usually through the internet. Daily observations and recent academic traditions suggest that some learning occurs beyond the confines of the individual mind. Research has shown that acquiring and learning languages is in nature socially constructed and interactive [1]. Language learning is interwoven with cultural interaction and "mediated by linguistic and other symbolic activity," according to the theory of language socialization [2]. From this point of view, it makes a lot of sense to use technologies that facilitate communication and

C. Troussas and M. Virvou, *Advances in Social Networking-based Learning*, Intelligent Systems Reference Library 181, https://doi.org/10.1007/978-3-030-39130-0_2

connection, especially social media applications and programs. Due to the different avenues in which new social media have been created, language learners are able to enhance their language skills. Social media allows the learner to participate in real-time, relevant online conversations and practice the target language with or without the help of an experienced teacher on his or her side.

The method of Social Media Language Learning (SMLL) consists of applying interactive social media channels to language learning, which in turn will enable the student to develop communication skills while using these social networks and advance in language learning.

The method provides the learner with the opportunity to participate in actual, real-time, relevant online conversations and practice target language by his or her side with the help of an experienced teacher.

The Communicative Language Teaching (CLT) method provided the basis for the development of the SMLL method, both emphasizing the importance of teaching with the objective of developing functional language knowledge within a wide range of contexts. Perfect grammar and pronunciation are not essential to the process, but rather focus on the student's communication skills and the ability to understand and understand him/herself.

The Social Media Language Learning is based on three principles:

1. Significance in the target language of live and actual communication through interaction and updated understanding and production of content based on social media channels.
2. Personal experience and interests of students play a defining role in learning, enabling relevant language use during and between classes with teacher and virtual community active participation.
3. In terms of editing, strategy, conceptualization, business insight, etc., the promotion of social media communication skills at the same time as language learning takes place.

Therefore, as all of them will result in learning, the learner is invited to engage as much as possible in activities that require the use of language. Communication both in-class and out-of-class is equally important. It combines the advantages of another method, known as Blended Learning, which allows students to learn autonomously whenever and wherever they want, with all the necessary materials available online, while at the same time having the support of an experienced teacher who facilitates the process and provides a professional and live explanation of the subjects at hand.

On-site classes are intertwined with the teacher with ongoing online conversations with other relevant individuals. Learning is considered a constant, ever-flowing, indivisible part of everyday life, making it part of the target language.

Social e-learning reflects many different social networking services features, such as Facebook. In addition, due to several beneficial features such as either enabling peer feedback and collaboration or interactivity and active participation, they can be highly regarded as an educational tool. They can improve informal learning and support social connections within learner groups and with those involved in learning support. Adopting a platform with social features can offer:

- Social influence: Due to the social character of these platforms, students can maintain communication or meet new friends with classmates. This fact therefore emphasizes the perception of social influence as an important factor in the decision by people to participate in social e-learning.
- Familiarization: Due to similar to user interfaces of widely used and commonly accepted social network sites (e.g. Facebook), the ease of using such platforms is accentuated.
- Cooperation: With the use of such platforms, the idea of collaborative learning can certainly be underlined. This allows students to exchange ideas, to help their peers and to collaborate to improve education.
- Usefulness: E-learning platforms that possess social properties can improve the productivity of individuals. In addition, different possibilities such as information sharing, collaboration and entertainment influence their adoption.
- Knowledge sharing: The exchange of resources, documents and useful information about the curriculum being taught is a key element of the educational use of Facebook. Moreover, they can also provide multimedia sharing so that students can offer audio, video, images and other material pertaining to their curriculum to their peers.
- Peer feedback: It is important to enable communication between users/students. As such, they gain knowledge about significant information about the curriculum being shared by others.

2.2 Related Literature for Social e-Learning

This section presents the related scientific literature for social e-learning systems using a novel ISO-based framework.

2.2.1 Methodology and Model Used

The literature review that is presented and discussed in this paper proceeded from a searching study of relevant papers being published in the last few years. The main criterion for a paper to be listed in the literature review was the presented e-learning system to be implemented with a social networking perspective or to be embedded in/developed using an existing social networking site. Moreover, the search engine used in this research was the Scopus, selecting articles published in qualitative research journals or papers presented at significant international conferences. Scopus was preferred since it is the largest abstract and citation database of peer-reviewed literature and it is considered as one of the most valid search engine for research papers.[1] Another criterion for the inclusion of papers was the system to

[1] https://www.scopus.com/.

be tested by their respective authors, as the evaluation was based on their system attributes description and testing results. Towards a qualitative review of the systems, a multicriteria framework was used and is presented below.

After the review of social media-based learning systems and software quality models, a quality analysis of selected systems was conducted using the proposed approach. To this end, the evaluation is relied on the system description and the testing results of their creators, as reported in their papers. The results of the evaluation have been tabulated and a comparative discussion has been conducted. Figure 2.1 illustrates the research methodology used in this paper.

2.2.2 Selected Systems in the Review

The current paper focuses on the evaluation of innovative educational systems that adopt social media and networking technologies. As this research area is in its infancy and growing day by day, the development of such systems is limited. Thus, after an extensive search of literature, the number of forty-one papers has been chosen, in which applications have been developed since 2010 to present. Moreover, they include a system testing section, essential for the evaluation.

With regard to papers' publication type, 65.85% of the selected systems have been published in qualitative research journals, and the rest ones have been presented at significant international conferences and have been published either as lecture notes or conference paper. Moreover, in 56.1% of the papers, the authors have developed an entire system with social networking and e-learning features, whereas in the rest papers the systems have been developed using well-known Web 2.0 technologies and LMS/CMS. In particular, almost halves of such systems exploit the capabilities of Facebook, the most popular social networking site, in order to establish a social e-learning application. Other Web 2.0 tools that have been used in the selected papers are Twitter—the most famous social networking microblogging site, Elgg—an open source social networking engine for developing social environments, Diigo—a collaboratively social annotation tool, Edu 2.0—a powerful e-learning platform with LMS and social networking features. In addition, there are systems implemented in Moodle—an open-source course management system, and Drupal—an open-source content management system. Finally, there are some cases where the system developed by the researches is related to Web 2.0 tools, either as Moodle plug-ins or Facebook apps.

Figures 2.2 and 2.3 show the statistics of the evaluated systems regarding the publication type and the platforms used for their development.

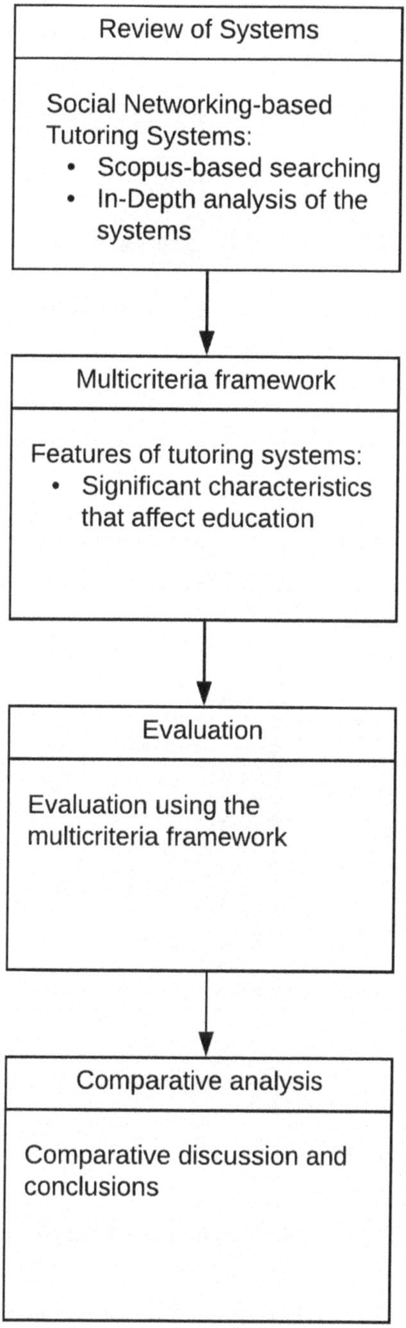

Fig. 2.1 Research methodology

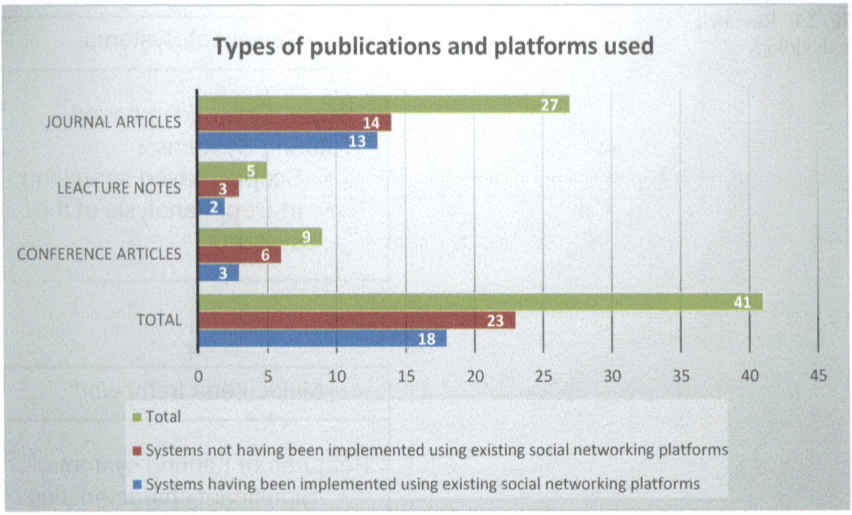

Fig. 2.2 Statistics of the evaluated systems regarding the publication type and platform used

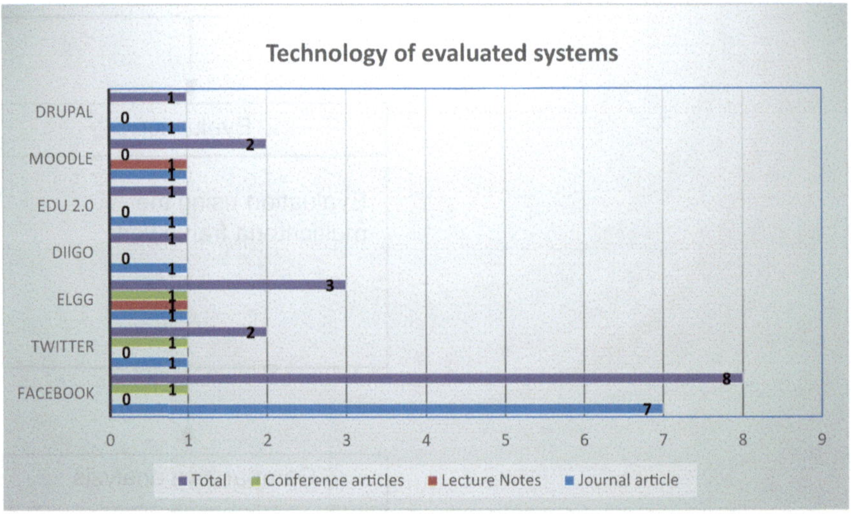

Fig. 2.3 Statistics of the evaluated systems regarding the technology used (Web 2.0/LMS/CMS)

2.2.3 Multicriteria Framework for Social e-Learning Systems

The development of Web 2.0 technologies and the spread of social media have dramatically changed the range and capacity of the web services delivered in general and in education in particular. There has been a wide range of social media-based learning systems. It is crucial for all benefits of e-learning and social media technology that high-quality systems are provided. However, for the quality of such systems, there is no standard assessment model. To this end, we use a multicriteria framework for the evaluation of the systems.

The multicriteria framework which was used for the review includes the following characteristics:

1. *Domain knowledge distribution Content delivery*: the system delivers the domain knowledge to learners.

Education is provided through Facebook posts where any type of file (text, video, image, etc.) can be attached [3–10]. Conversations can be followed through the use of Twitter by the corresponding Twitter hashtags [8, 11]. Diigo allows users to mark, comment on and share their annotations with others on websites or documents [12]. The tutors are easily able to add content to the class using Moodle [13, 14]. Along with components like posts, file shares or bookmarks, Elga enables tutors to deliver the course material [15–17]. The user can generate their content via their blogs or groups in myCourse [18] other than through the learning content provided by the platform. Omega [19] provides a formal teacher course and content sharing between students that can be assessed by other students in order to promote useful material on the same basis. Likewise, the function of uploading and downloading material is also embodied by Book2U [20], and SaxEx [21]. The Class is a social network with an integrated STI, organized in a tree structure into chapters and topics, Fermat [22]. SoACo [23] transforms content in education-supportive support systems from social networks such as Facebook and Twitter. Veeramanickam and Radhika [24] finally proposed a smart e-learning system with LMS and SNS featuring, while Rožac et al. [25], via the Facebook application, integrated Coome LMS with the Facebook platform.

2. *Depiction of development of learners' knowledge level Management of student records & tracking students' progress*: the system keeps the student's records such as grade, error proneness or the specific part of the his/her studies.

Few systems can monitor students' development by focusing on the social aspect of education [22, 26, 27]. The monitoring of student actions is a difficult task on platforms such as Facebook, Twitter and Diigo because there is no log file and the filter option for the post/comments is considerably restricted. The mass of information uploaded makes it difficult for the students to browse through comments and the holistic view of their activities. Moodle provides an ingraded progress tracking system, including qualifications, course and activities completion reports, etc. as a high-powerful learning platform [13, 14].

3. *Cooperation and Information exchange Communication & collaboration*: the ability for students or their tutors to cooperate with peers.

Facebook offers students the ability to communicate and collaborate with other students and teachers in a synchronous or asynchronous manner through postings, comments and private chat. In the meantime, students on Twitter interact only with tweets and tweets with others [11]. Diigo enables users to share annotations and discuss comments with others [12]. A variety of tools for communication and collaboration between students such as the forum, the chat, the blog, the sharing of bookmarks and others (including Elgg-based systems) [15, 17] and Drupal-based SNAP [28] are also available to Edu 2.0 [29], and Moodle [13, 30]. Communication and cooperation are achieved through the MOOC platform, twitter, and google + in the García-Peñalvo et al. [14] system, which includes specific hashtags. SaxEx [21], Topolor [27], Book2U [20], PREBOX [31] and myCourse [18] are also used to implement social characteristics for comment, communication, rating/liking, and so forth. Moreover, the collaborative functionality of learning and networking systems such as weSPOT [32] and Edil-learning [33] and ColeSN [34] is accentuated.

4. *Learner clustering Organizing students into groups*: the possibility of creating groups in order to enable students to work on shared projects.

Research efforts in which students can participate in various groups and exchange their views, information etc., with other people are available in: Gao [12], the Diigo-bases system; Chunyan et al. [29], Edu 2.0, SocialWire [17], Elgg-bases one; 3]. Chuang et al. [35] proposed a student-based grouping method for better learning outcomes, based on friendship, test grades, pairing algorithms and evaluations, while Arndt and Guercio [36] proposed a way of providing common learning experiences based on their connectivity within social networks. Within this context, Hsu et al. [26], based on student knowledge, implemented a grouping system on Facebook. Lintend [37] and MyLearnSpace [38] are other systems which enable students to enter groups according to their interests in the future.

5. *Test delivery Conducting assessments & maintaining records of assessments*: the system's ability to deliver different types of assessments to test knowledge of students (e.g. multiple choice exercises or gap filling exercises, etc.). In addition, it is about maintaining the evaluation records of each student's model.

In several systems where there is not support of test delivery, the tutors should either use other tools of web 2.0 in testing or upload testing assignments as files and manage their results manually [9–12]. This is because Facebook, Twitter and Diigo have no evaluation tool. Tools such as Moodle and Edu 2.0 provide testing and online grading components [13, 14, 29]. A video portfolio, consisting of all kinds of materials produced by the students, the reputation and the gradebook, was implemented in SocialWire [17] an Elgg-based system for the purpose of the quizzes and exams to enable the creation of traditional student testing and automatic grading. With respect to the other systems, only few provide an integrated evaluation system, including a quiz [22, 24, 27].

6. *Learning attainment Learning outcome*: the students' learning attainment resulting from the instructive process is analyzed by the system.

Based upon student activities on the Moodle platform and meaningful learning features, Mansur and Yusof [13] classifies the learner behavior into active, constructive, and intentional. García-Peñalvo et al. [14] deploys Moodle for the collection of social networking information shared by students on the MOOC learning process platform. The SocialWire Proposal [17] applies the following headings to assess the achievement of all learning activities. SaxEx [21] is also adopting a badge system which rewards students on the basis of the answers, likes, places and comments of the triggered questions. Similarly, on reaching certain objectives in inquiries, weSPOT [32] defines badges. While Rampun and Barker [39] uses renowned points to motivate users to become more involved, by uploading, postings, discussion etc. Finally, Fermat uses the cognitive ACT-R theory [22].

7. *Software for creating e-learning content Authoring tool*: the system offers instructors a tool to easily create professional, engaging and interactive learning content, since they are not capable of programming.

A basic authoring instrument for administrators to manage group members and group settings like confidential, posts etc. is found in Facebook groupings [3, 4, 9, 40]. On the other hand, teachers and students can use the appropriate options in a graphical environment for applications developed using CMS like Moodle. There are also few systems which provide a tool for authoring. Hsu et al. [26] have developed a collaborative learning Facebook application that includes an instructor management interface, for example. Student data and student groups can be managed by SaxEx [21] teachers and location-based issues. There is an expert group in S-LCMS [41] that creates learning articles using appropriate components, such as content generation, import, export, publishing etc.

8. *Delivery of material from external sources Access content from and provide content to digital libraries & other e-learning systems*: the system can deliver material from external sources (material of electronic libraries or other digital tutoring systems) to users.

Through Web 2.0. Tools, included in this work, users can easily share and upload material from other resources. (for example, [4, 16], and in the majority of other systems (e.g. [14, 16, 28, 23, 24, 29])).

9. *System's support to erroneous interaction System response (to invalid input data)*: the system supports learners when facing problems in the interaction with the system by providing messages and help.

This is an important feature to be included by any tutoring system, especially a learning environment, as students may not be sufficiently familiar with computers and need guidance for learning. The papers used in this survey do not mention this feature however because they concentrate on their systems' innovative capabilities.

10. *Error routine Error management/handling*: the system provides a routine (namely a series of actions) in cases students make a mistake while interacting with the system.

This ability is very important in order to develop stable systems. However, solely their innovations are described in the articles presenting the evaluated systems.

11. *Easiness to use Consistency of layout (user friendliness)*: the system has the same layout concerning important actions made by users. For example, help buttons in all system forms should be in the same place.

Most systems have been implemented based on interface design principles. The graphical user interface (GUI), therefore, is freindly, easy to use, and well-organized.

12. *Simple input of data Clear prompts for input*: the system provides a clear understanding to students when they are asked to input data, such as a questionnaire completion.

The clear understanding for input data underlines the existence of a neat user interface by providing information for the correct use of controls. Tools from Web 2.0 take this function into account on their interfaces, while most of the systems used in this work did not refer to it in the paper. For example, Fermat provides students with the correct teaching method and personalized assistance when it is difficult to respond to test questions [22]. In addition, appropriate speeds in its input fields are observed in Topolor [27].

13. *Messages supporting users Help messages*: The system assists and protects learners from error making.

In order to facilitate system navigation and use of all system capabilities, it is important to have the software provide a well-designed help system. Unfortunately, this feature is not mentioned for analyzing additional capabilities in the selected systems.

14. *Problems during the interaction with the system Difficulty when learning to operate the system*: upon their first interaction with the system, students may face problems pertaining to the operation.

Generally, there is no any reference concerning learners' problem while interacting with the system except from SaxEx [21].

15. *Proper presentation of information Organized information and sequence of screens*: learners' information is properly organized and presented in a systematic way.

Many Web 2.0 tools, such as Facebook, Twitter, blogs and so on, mainly arrange their material based on chronology or tags it has uploaded. Diigo is in the meantime able to manage it in directories. Edu 2.0 and Moodle provide an efficient and efficient way of organizing the lessons, as learning management systems. In Elgg, different plugins that support the desired functionality allow this feature to be achieved. Finally, during his/her interaction, SaxEx [21] shows the questions according to the location of the student.

16. *User Interface likability Pleasantness/attractiveness of system interface*: there are system characteristics that increase the user's enjoyment and satisfaction, such as color and graphic design nature.

The most well-known platforms are Facebook and Twitter, which are widely used by all ages. Their interface therefore familiarizes learners. They are regarded as pleasant and easy to use environments for learners in conjunction with their easiness of use. Every Web 2.0 tool has a simple and usable interface, and developers are generally able to use systems that are effective, efficient and satisfying.

17. *Individualization Personalization*: the system is tailored and adapted to the interests, preferences and needs of individual users.

Edu 2.0 platforms [29] and Elgg platforms [15] allow users to adjust or remove the option to configure their dashboards. S-LCMS [41] offers personalized learning by means of educational objects based on various study and cognition styles. The research at García-Peñalvo et al. [14] is a follow-up process for students' social networking conversations, so that they are able to use this knowledge to adapt MOOC content. Another social network, Zamna [42], adapts its content related to the identified student learning style based on Felder-Silverman model. In addition, Fermat [22] is adapted based on cognitive aspects and students' recognized emotion. Whereas Chuang et al. (2012) implements an adaptive system by providing adaptive grouping and adaptive tests related to groups.

18. *Advice on learning process System recommendations*: the system can offer students appropriate advice or learning material to the direction of meeting their needs in a dynamic way.

Di Bitonto et al. [16], using Elgg, proposed a suggestive approach that would suggest study objects, users and discussion groups related to the needs of learners and adjust search results according to the interests of the learner. This method was implemented with user defined tags and an algorithm for clustering. In Topolor [27], another approach is used, where content and peer recommendation are provided through the module and Q&A center. In addition Omega [19] offers suggestions on adaptive material and filters, using user-based algorithms of closest neighbour, based on the utility rating of community-and student mental efforts.

2.3 Comparative Discussion

First, a comparison has been conducted on the prevailing learning approaches through SNS. As already mentioned, the comparison has been made using a multi-criteria framework to produce qualitative results. As shown in Fig. 2.4, "Domain Knowledge distribution" is the most commonly used e-learning feature. "Domain Knowledge distribution" is important because the way a student receives the learning material is involved.

Moreover, as shown in the chart of Fig. 2.4, "Cooperation and information exchange" is also commonly used. Collaborative learning supported by computers (CSCL) is a pedagogical approach in which learning takes place via a computer or the Internet via social interaction. Since SNSs offer different ways of asynchronous and

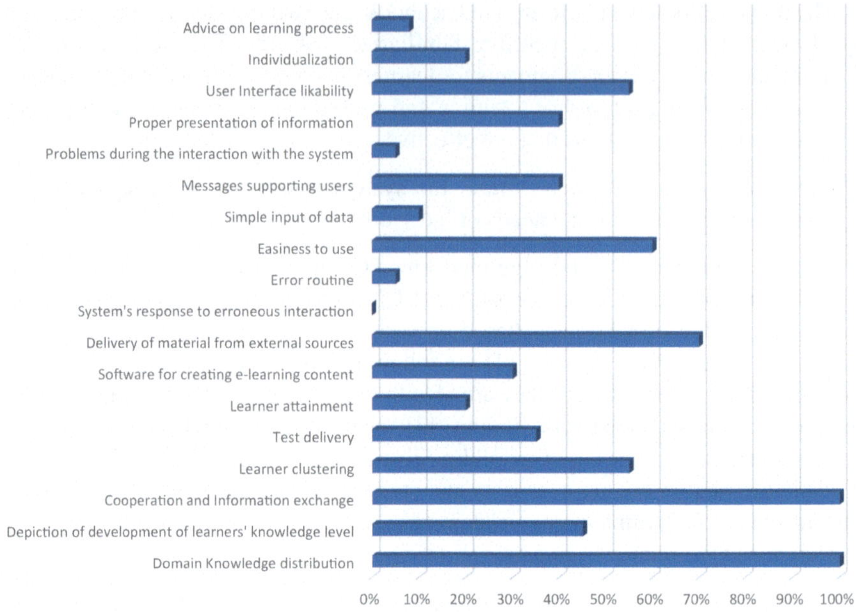

Fig. 2.4 Percentages of characteristics of the multicriteria framework

synchronous communication with the students, collaboration is the most frequently used characteristic in such systems from 2010 until now.

In addition, researchers preferred to conduct "Learner clustering" between 2013 and 2015. The students' participation in group work is a key component of this feature. When students work together, they can express their own thoughts, listen to the point of view of their colleagues and thus remain at the heart of the teaching process.

As illustrated in Fig. 2.4, "User interface likability" is highly taken into consideration. They play an important role in education, because students need an attractive and consistent arrangement in order to focus all their attention on learning.

The "individualization" of students must be the fundamental pillar of the educational process in traditional e-learning systems. Students are specifically placed at the tutoring facility and they are adapted to every learning object and function. But the percentage of individualization as an ability in social e-learning systems is quite low. Lastly, the social aspect of education is the focus of these systems: communication, cooperation and grouping. This happens because these systems can be seen as a growing problem in science literature. They include recently widely used modules and features such as supporting customization, adaptability and advice.

Another important note is that most of these systems do not adopt a model or theory of learning styles. The support for a model or theory of a style of learning is important because identifying the learning method for students is key to introducing techniques and strategies for sequencing the curriculum and evaluation methods.

As far as the testing of student knowledge is concerned, it is found that systems provided simple means such as static and predefined testing and non-dynamic student error correction. But it is crucial for assessing the achievement of the learning objectives to use an assessment tool that adapts their activity to the student model and supports error diagnostics.

Summarizing, characteristics such as Domain knowledge distribution, Cooperation and information exchange and Delivery of material from external sources are have been taken into consideration in a high percentage for the construction of social networking learning systems. A considerably lower percentage in appearance in the construction of social networking learning systems is held by characteristics such as Depiction of development of learners' knowledge level, Learner clustering Easiness to use and User interface Likability. An even lower percentage is held by characteristics such as Test Delivery, Learning attainment, Software for creating e-learning content, Proper presentation of information and Individualization. Finally, characteristics such as System's support to erroneous interaction, Error routine, Simple input of data, Messages supporting users, Problems during the interaction with the system and Advice on learning process present a very low percentage to zero percentage for being taken into account for the implementation of the aforementioned systems.

References

1. McClanahan, L.: Training using technology in the adult ESL classroom. J Adult Educ. **43**(1), 22–27 (2014)
2. Reinhardt, J., Zander, V.: Social networking in an intensive English program classroom: a language socialization perspective. CALICO J. **28**(2), 326–344 (2011)
3. Stankov, S.V., Glavinic, V., Krpan, D.: Group modeling in social learning environments. Int. J. Distance Educ. Technol. **10**(2), 39–56 (2012)
4. Asterhan, C.S., Rosenberg, H.: The promise, reality and dilemmas of secondary school teacher–student interactions in Facebook: The teacher perspective. Comput. Educ. **85**, 134–148 (2015)
5. Milošević, I., Živković, D., Arsić, S., Manasijević, D.: Facebook as virtual classroom–social networking in learning and teaching among Serbian students. Telematics Inform. **32**(4), 576–585 (2015)
6. Sharma, S.K., Joshi, A., Sharma, H.: A multi-analytical approach to predict the Facebook usage in higher education. Comput. Hum. Behav. **55**, 340–353 (2016)
7. Güler, K.: Social media-based learning in the design studio: a comparative study. Comput. Educ. **87**, 192–203 (2015)
8. Manca, A., et al.: Integrating twitter into an undergraduate medical curriculum: lessons for the future. In: Proceedings of the European Conference on e-Learning, ECEL, pp. 330–338 (2014)
9. Meishar-Tal, H., Kurtz, G., Pieterse, E.: Facebook groups as LMS: a case study. Int. Rev. Res. Open Distance Learn. **13**(4), 33–48 (2012)
10. Raud, Z.V., Vodovozov, V., Lehtla, T.: Enhancing students' activity in electrical engineering through web and social networks. In: IEEE Global Engineering Education Conference, EDUCON (2012)
11. Junco, R., Heiberger, G., Loken, E.: The effect of twitter on college student engagement and grades. J. Comput. Assist. Learn. **27**(2), 119–132 (2011)
12. Gao, F.: A case study of using a social annotation tool to support collaboratively learning. Internet High. Educ. **17**(1), 76–83 (2013)

13. Firdausiah Mansur, A.B., Yusof, N.: Social learning network analysis model to identify learning patterns using ontology clustering techniques and meaningful learning. Comput. Educ. **63**, 73–86 (2013)
14. García-Peñalvo, F.J. et al.: Evolution of the conversation and knowledge acquisition in social networks related to a MOOC course. In: Lecture Notes in Computer Science (including subseries Lecture Notes in Artificial Intelligence and Lecture Notes in Bioinformatics), 9192, pp. 470–481 (2015)
15. Veletsianos, G., Navarrete, C.C.: Online social networks as formal learning environments: Learner experiences and activities. Int. Rev. Res. Open Distance Learn. **13**(1), 144–166 (2012)
16. Di Bitonto, P., Roselli, T., Rossano, V.: Recommendation in e-learning social networks. In: Lecture Notes in Computer Science (including subseries Lecture Notes in Artificial Intelligence and Lecture Notes in Bioinformatics), 7048, pp. 327–332 (2011)
17. Sousa-Vieira, M.E., et al.: SocialWire: Social software for informal learning. In: Proceedings of the 2nd International Workshop on Interaction Design in Educational Environments (IDEE-2013), pp. 133–145 (2013)
18. Giouvanakis, T.: Mycourse: an educational content management platform with social networking characteristics. Int. J. Emerg. Technol. Learn. **5**(4), 17–25 (2010)
19. Dominoni, M., Riva, G., Pinardi, S.: Omega network: an adaptive approach to social learning. In: Proceedings of the 2010 10th International Conference on Intelligent Systems Design and Applications, ISDA'10, pp. 953–958 (2010)
20. Balakrishnan, V.: Using social networks to enhance teaching and learning experiences in higher learning institutions. Innov. Educ. Teach. Int. **51**(6), 595–606 (2014)
21. Boticki, I., et al.: Usage of a mobile social learning platform with virtual badges in a primary school. Comput. Educ. **86**, 120–136 (2015)
22. Zatarain-Cabada, R., Barrón-Estrada, M.L.: Affective tutoring systems in a learning social network. Int. J. Contin. Eng. Educ. Life Learn. **23**(1), 91–99 (2013)
23. Kim, K., Moon, N.: Content modeling based on social network community activity. J. Inf. Process. Syst. **10**(2), 271–282 (2014)
24. Veeramanickam, M.R.M., Radhika, N.: A smart e-learning system for social networking. Int. J. Electr. Comput. Eng. **4**(3), 447–455 (2014)
25. Rožac, J., Ballester, J.V., Buendía, F.: A model-based approach to integrate e-learning platforms and social networks using a service-oriented framework. J. Res. Pract. Inf. Technol. **44**(3), 247–265 (2012)
26. Hsu, C., et al.: The development of an adaptive group composition system on facebook for collaborative learning using an artificial bee colony algorithm. Appl. Math. Inf. Sci. **8**(1), 157–164 (2014)
27. Shi, L., et al.: A social personalized adaptive e-learning environment: a case study in Topolor. IADIS Int. J. WWW/Internet **11**(3), 1–17 (2013)
28. Kirkwood, K.: The SNAP platform: Social networking for academic purposes. Campus-Wide Inf. Syst. **27**(3), 118–126 (2010)
29. Chunyan, L., Haitao, C., Guolin, L.: The effect of Web 2.0 on learning management system. Int. J. Multimed. Ubiquitous Eng. **9**(10), 67–78 (2014)
30. Garmendía, A., Cobos, R.: Towards the extension of a LMS with social media services. In: Lecture Notes in Computer Science (including subseries Lecture Notes in Artificial Intelligence and Lecture Notes in Bioinformatics), 8091, pp. 67–72 (2013)
31. Rodrigues, J.J.P.C., Sabino, F.M.R., Zhou, L.: Enhancing e-learning experience with online social networks. IET Commun. **5**(8), 1147–1154 (2011)
32. Mikroyannidis, A., et al.: WeSPOT: a personal and social approach to inquiry-based learning. J. Univers. Comput. Sci. **19**(14), 2093–2111 (2013)
33. Longo, F., Santoro, S., Esposito, A., Tarricone, L., Zappatore, M.: Integrating social web and e-learning to enhance cooperation in the building sector. In: Proceedings of 2014 International Conference on Interactive Collaborative Learning, ICL 2014, pp. 324–329 (2014)

34. Caballe, S., Britch, D., Barolli, L., Xhafa, F.: A methodological approach to provide effective web-based training by using collaborative learning and social networks. In: Proceedings-2014 8th International Conference on Complex, Intelligent and Software Intensive Systems, CISIS 2014, pp. 64–71 (2014)
35. Chuang, P.-J., Chiang, M.-C., Yang, C.-S., Tsai, C.-W.: Social networks-based adaptive pairing strategy for cooperative learning. Educ. Technol. Soc. **15**(3), 226–239 (2012)
36. Arndt, T., Guercio, A.: Social network-based course material transformations for a personalized and shared ubiquitous e-learning experience. In: UBICOMM 2011—5th International Conference on Mobile Ubiquitous Computing, Systems, Services and Technologies; PECES 2011—3rd International Workshop on Pervasive Computing in Embedded Systems, pp. 218–222 (2011)
37. Popescu, E., Ghita, D.: Using social networking services to support learning. In: Lecture Notes in Computer Science (including subseries Lecture Notes in Artificial Intelligence and Lecture Notes in Bioinformatics), 8167, pp. 184–193 (2013)
38. Hubwieser, P., Mühling, A.: A social network for learning: supporting collaborative learning based on the ontology for educational knowledge. In: Proceedings of the International Conference on Knowledge Management and Information Sharing (KMIS), pp. 298–301 (2012)
39. Rampun, A., Barker, T.: The design and development of an eLearning system based on social networking. In: Proceedings of the European Conference on Games-Based Learning, pp. 646–657 (2011)
40. Lin, P.: Analyzing knowledge dimensions and cognitive process of a project-based online discussion instructional activity using facebook in an adult and continuing education course. Comput. Educ. **60**(1), 110–121 (2013)
41. Kim, K., Moon, N.: Designing a social learning content management system based on learning objects. Multimed. Tools Appl. **64**(2), 423–427 (2013)
42. Zatarain-Cabada, R., Barrón-Estrada, M.L., Angulo, V.P., García, A.J., García, C.A.R.: A learning social network with recognition of learning styles using neural networks. In: Lecture Notes in Computer Science (including subseries Lecture Notes in Artificial Intelligence and Lecture Notes in Bioinformatics), 6256, pp. 199–209 (2010)

Chapter 3
Intelligent, Adaptive and Social e-Learning in POLYGLOT

Abstract Intelligence and adaptivity are core characteristics of sophisticated e-learning systems which in many cases have social networking functionalities. They involve the development of modeling, diagnostic, feedback and reasoning techniques towards creating a student-centered learning environment. Therefore, there have been systems that combine all of the above technologies for tutoring specific domains. However, there have also been domain independent tutoring platforms with social characteristics that serve as learning management systems within the technology of social media. In this chapter, we give a short presentation of the concepts of domain independent mechanisms, including student modeling techniques, as well as of learners' characteristics for building efficient student models. The described concepts are used in the following chapters for the implementation of the social networking-based language learning application, called POLYGLOT. In addition, this chapter presents well-known web-based platforms that have been used for social e-learning.

3.1 Intelligent Tutoring Systems (ITSs)

An Intelligent Tutoring System (ITS) may be described neatly as applying Artificial Intelligence (AI) to educational contexts. ITSs' intelligence lies in adapting the tutoring process to the needs of learners, which means offering different tutoring to each individual student. The rapid development of high and new technology in Computer Assisted Instruction (CAI) has opened new horizons over the past decades. Indeed, computer intrusion affected the so-called ITS architectures. ITSs are broadly defined computer systems that incorporate components of AI. Such systems can be designed to provide learners with immediate and customized instruction or feedback [1], usually as low as possible or even without a human teacher's intervention. More specifically, ITSs are attempting to emulate human tutors' approaches and language in order to support real-time or on-demand instructional interactions, as individual learners exactly need. ITSs are defined as "computer-based tutoring systems that incorporate instructional content models designating what to teach and teaching strategies designating teaching methods" [2]. In ITS, the sequencing of learning content is customized to avoid a cognitive mismatch that can be caused by providing

C. Troussas and M. Virvou, *Advances in Social Networking-based Learning*, Intelligent Systems Reference Library 181, https://doi.org/10.1007/978-3-030-39130-0_3

low performers with difficult learning content and providing high performers with non-challenging tasks. Adapting the learning content based on the needs of the student and personalizing the student's learning allows ITS to work with students of different skills.

An ITS' overall objective is to solve the problem of over-dependence between students and teachers in the direction of providing quality education. It aims to provide each and every student with access to high-quality education, thereby reforming the entire education system. ITSs' goal is to track the progress of learners, tailor feedback and suggestions to their needs [3]. The ITS can make inferences about its strengths and weaknesses by holding information about the performance of a particular student, and may even suggest additional work [4]. ITS implementation includes computational mechanisms and representations of knowledge in the fields of AI, computer linguistics, and cognitive science. As such, there is a close relationship between smart tutoring, cognitive learning theories and design (Fig. 3.1); and ongoing research is underway to enhance ITS' effectiveness.

ITSs should involve several features as follows [5]:

- To enable tutoring of every task for people with disabilities, giving more autonomy in working environments.
- To have a multimodal task management system associated with each individualized profile to integrate data from different sources (speech, images, videos, and text).
- To be integrated into a mobile platform, i.e. a mobile telephone or PDA (Personal digital assistant).
- To contain a multimedia interface that has to be friendly, reliable, flexible, and ergonomically adapted.
- To integrate a human emotional predictive management in order to prevent risk, emergency and blockage situations that can damage these people and interfere with their integration into working and social environments.

Fig. 3.1 Domain of ITS

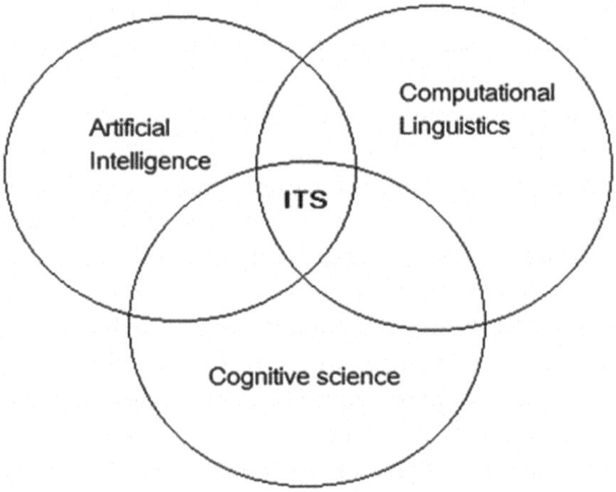

- To be entirely configurable by stakeholders without technological knowledge in order to enable an easy and flexible access.
- To show the capability of exporting the system to other populations, i.e. the elderly.

3.1.1 Architecture of ITS

The architecture of ITS consists of four basic and interrelated modules, namely the Learning Content, the Student Model and the Adaptation Engine and the User Interface [6]. The generic architecture of the ITS is shown in the Fig. 3.2.

3.1.1.1 Learning Content

ITS' learning content is a set of domain topics. Such topics are divided into learning units that support a particular concept or fact tutoring. The system database holds the misconceptions of possible students and common erroneous answers for each learning unit. The learning units may have the form of explanations, instances, intimations, tests, exams, and may be used to educate, present, or evaluate the students. The most important function of this model is to provide a structure to represent the knowledge of the user domain.

This value can be expressed quantitatively, qualitatively or in probabilistic form.

3.1.1.2 Student Model

The student model contains several student information, e.g. their level of education, previous knowledge and background in order to provide a student-centered learning experience [7, 8, and 9]. This model also stores other types of student information, such as:

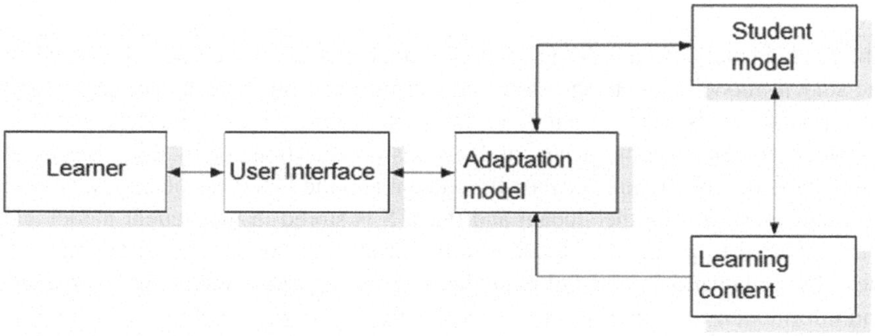

Fig. 3.2 Generic architecture of ITS

- The skills, the goals and the plans of students.
- Student's performance such as topic performance and number of questions correctly answered per session.
- Learning characteristics such as the learning rate, the student's preferences and learning styles.
- Affective states such as engagement, boredom, frustration and confusion.

3.1.1.3 Adaptation Model

The adaptation engine is a technique or algorithmic approach for adapting the student's learning content based on his/her input through the user interface (e.g. answering questions) and the student model information [10, 11]. An ITS adapts the learning content to the preferences of the learner such as:

- The learner's level of ability, such as, "novice" or "intermediate" or "expert" [12].
- The learner's knowledge, such as, previous knowledge of learning content [13].
- Learning styles such as "visual", "audio", and "interactive" [14].

3.1.1.4 User Interface and Log File

The user interface provides the student with the learning content and accepts the responses of the students to the ITS questions [15, 16]. The learning content can be provided as text, voice, simulation or even interactive games based on the nature of the ITS. A mobile device (Tablet, Mobile, Laptop) or a desktop can be the user interface. The interaction of the students with the ITS is captured in the log file, such as answering questions, number of attempts and time taken for different activities (responding, reading, etc.). The log file is used as the student model's input.

3.1.2 Function of ITSs

The fundamental function in ITSs is to initialize a student upon registration when the student model collects and stores crucial information such as age, gender and background etc. Based on their level and preferences, the user interface supports the students. For example, when students answer questions or by other means of interaction, the interaction between the student and the ITS takes place. Every type of interaction between the student and the ITS is stored in the student model and then analyzed to promote adaptation to the needs and preferences of the student. Next, ITS' adaptation model tailors students' learning content based on their request and information from their profile. For example, if a student performs an error in a quiz, the ITS can diagnose the reason for the error and support the student by giving him/her advice to overcome the misconception.

3.2 Intelligent Computer-Assisted Language Learning (ICALL)

Computer-assisted language learning (CALL) can be defined briefly as "searching and studying computer applications in language learning and teaching" [17]. CALL encompasses a wide range of information and communications technology applications and approaches to the direction of foreign language teaching and learning, from the "traditional" drill and practice programs that characterized CALL in the 1960s and 1970s to more recent manifestations of CALL, e.g. as used in a virtual learning environment and distance learning based on the Web. It also covers corporate and concordance use, interactive whiteboards, computer-mediated communication (CMC), virtual world language learning, and mobile-assisted language learning (MALL).

Before CALL, the term CALI (computer-assisted language instruction) was used as a subset of the general term CAI (computer-assisted instruction) reflecting its origin. However, CALI fell out of favor among language teachers as it seemed to imply an (instructional) teacher-centered approach, whereas language teachers are more inclined to prefer a student-centered approach, focusing on learning rather than instruction. In the early 1980s, CALL began replacing CALI [18] and is now incorporated into the names of the growing number of professional associations around the world.

CALL's current philosophy places great emphasis on materials that are student-centered and allow students to study alone. Such materials may be structured or unstructured, but normally incorporate two important characteristics: interactive learning and individualized learning. CALL is essentially a tool that helps instructors facilitate the process of language learning. It can be used to improve what has already been taught in the traditional classroom or as a remedial tool to help students who need additional support.

The design of CALL materials generally takes into account the principles of language pedagogy and methodology that can be derived from different theories of learning (e.g. behavioral, cognitive, constructivist) and second-language learning theories such as the hypothesis of Stephen Krashen.

Several attempts were made in the 1980s and 1990s to establish a typology of CALL. Davies and Higgins [19], Jones and Fortescue [20], Hardisty and Windeatt [21], and Levy [17] have identified a wide range of different types of CALL programs. These included gap-filling and closing programs, multiple-choice programs, free-format (text-entry) programs, adventures and simulations, action labyrinths, sentence-reordering programs, exploratory programs—and "total cloze," a type of program in which the learner has to reconstruct a whole text. In modernized versions, most of these early programs still exist.

CALL has become increasingly difficult to categorize since the 1990s as it now extends to blogging, wikis, social networking, podcasting, Web 2.0 applications, virtual world language learning and interactive whiteboards [22].

A different approach was taken by Warschauer [23] and Warschauer and Healey [24]. Instead of focusing on CALL typology, three historical phases of ICALL were identified, classified according to their pedagogical and methodological approaches:

- Behavioristic CALL: conceived in the 1950s and implemented in the 1960s and 1970s.
- Communicative CALL: 1970s to 1980s.
- Integrative CALL: embracing Multimedia and the Internet: 1990s.

The majority of CALL programs in the first phase of Warschauer and Healey [24], Behavioristic CALL (1960s–1970s), consisted of drill-and-practice materials in which the computer gave a stimulus and the learner gave an answer. At first, it was possible to do both only by text. The computer would analyze the input of students and provide feedback, and more sophisticated programs would react by branching to help screens and remedy activities to the mistakes of students. While such programs and their underlying pedagogy still exist today, most language teachers have rejected behavioral approaches to language learning, and computer technology's growing sophistication has led CALL to other possibilities.

The second phase described by the Communicative CALL Warschauer and Healey [24], is based on the communicative approach that became prominent in the late 1970s and 1980s [25]. In the communicative approach, the focus is on language use rather than language analysis, and grammar is implicitly taught rather than explicitly. It also allows the student language output to be original and flexible. The communicative approach coincided with the PC's arrival, which made computing much more widely available and resulted in the development of foreign language learning software being a burning issue. In this phase, the first CALL software continued to provide skill practice but not in a drill format—for example: paced reading, reconstruction of text, and language games—but the computer remained the tutor. Computers provided context for students to use the language in this phase, such as as asking for directions to a place, and programs not primarily designed for language learning were used to teach foreign languages. Criticisms of this approach include using the computer for more marginal purposes in an ad hoc and disconnected manner than for the central purposes of language instruction.

Starting in the 1990s, the third phase of CALL described by Warschauer and Healey [24], Integrative CALL, attempted to address criticisms of the communicative approach by integrating language skills teaching into tasks or projects to provide direction and coherence. It also coincided with the development of multimedia (text, graphics, sound and animation) technology as well as CMC technology. CALL saw a definitive shift from using the computer for drilling and tutorial purposes (the computer as a finite, authoritative basis for a particular task) to a medium for extending education beyond the classroom.

Multimedia CALL began with interactive laser videodisks showing simulations of situations where a key role was played by the learner. Warschauer [26] later renamed the CALL as Structural CALL and revised the three phases as follows:

- Structural CALL: 1970s to 1980s.
- Communicative CALL: 1980s to 1990s.
- Integrative CALL: 2000 onwards.

Bax [27] took issue with Warschauer and Haley [24] and Warschauer [26] and proposed these three phases:

- Restricted CALL—mainly behaviouristic: 1960s to 1980s.
- Open CALL—i.e. open in terms of feedback given to students, software types and the role of the teacher, and including simulations and games: 1980s to 2003.
- Integrated CALL—Bax [27] argued that at the time of writing language teachers were still in the Open CALL phase, as true integration could only be said to have been achieved when CALL had reached a state of "normalization", namely when using CALL was as normal as using a pen.

ICALL is concerned with presenting multiple challenges in all language learning dimensions. Such challenges include strategies for designing and implementing the use of AI in language acquisition tutoring systems. In terms of handling noisy situations, ICALL systems should primarily be able to improve the learning process. ICALL should also incorporate pedagogical or cognitive language theories that can support students in their efforts. The learning goals the students set should be clear. As such, ICALL systems should be able to distinguish and model each learning case to an appropriate degree of granularity. Therefore, students can have the potential to determine their progress towards language learning or their weaknesses.

Pedagogy must above all be given careful consideration when designing ICALL software, but ICALL software publishers tend to follow the latest trend, irrespective of their desirability. In addition, approaches to teaching foreign languages are constantly changing, going back to the more recent communicative approach and constructivism by means of the direct method, audio-lingualism and a variety of other approaches [28].

Designing and creating ICALL software is a very demanding task that calls for a variety of skills. A team of people usually manage major ICALL development projects:

- A subject specialist (also known as a content provider)—usually a language teacher—who is responsible for providing the content and pedagogical input. More than one subject specialist is required for larger ICALL projects.
- A programmer who is familiar with the chosen programming language or authoring tool.
- A graphic designer, to produce pictures and icons, and to advise on fonts, color, screen layout, etc.
- A professional photographer or. Graphic designers often have a background in photography too.
- A sound engineer and a video technician will be required if the package is to contain substantial amounts of sound and video.

- An instructional designer. Developing a CALL package is more than just putting a text book into a computer. An instructional designer will probably have a background in cognitive psychology and media technology, and will be able to advise the subject specialists in the team on the appropriate use of the chosen technology [29].

ICALL inherently supports the autonomy of learners, the end of the eight conditions referred to by Egbert et al. [30] as 'Conditions for optimal language learning environments.' Learner autonomy places firm control over the learner so that he/she decides on learning goals.

Authoring tool seems to be a powerful idea when designing ICALL software to produce a set of multiple-choice and gap-filling exercises using a simple authoring tool [31], but ICALL also has to do with creating and managing an environment that incorporates a constructivist and linguistic philosophy [32]. According to constructivist theory, learners are active participants in tasks where they "build" new knowledge from their previous experience. Learners also assume responsibility for their learning, and the teacher is not a knowledge provider but a facilitator. Whole language theory embraces constructivism and postulates that language learning moves from the whole to the part, instead of building sub-skills that lead to higher comprehension, speech, and writing skills. It also emphasizes the interrelated skills of understanding, speaking, reading, and writing, reinforcing one another in complex ways. Therefore, language acquisition is an active process in which the learner focuses on signs and meaning and makes smart guesses. Additional requirements are placed on teachers working in a technological environment that incorporates constructivist theories and language theories as a whole. Developing the professional skills of teachers must include new skills in both pedagogy and technical and management. The teacher has a key role to play in the issue of teacher facilitation in such an environment, but there may be a conflict between the goal of creating an atmosphere for learner independence and the natural sense of responsibility of the teacher. To avoid negative perceptions of learners, Stepp-Greany [32] points out that continuing to address their needs, especially those of low-capacity learners, is particularly important for the teacher.

3.3 User Modeling and Adaptivity

The basis for customization in computer-based educational applications is a student model. In any intelligent or adaptive tutoring system, it is a core component that represents many of the features of the student such as knowledge and individual traits [6]. Self [33] pointed out that student modeling is a process dedicated to representing several cognitive issues such as analyzing the performance of the student, isolating the underlying misconceptions, representing the goals and plans of the students, identifying prior and acquired knowledge, maintaining episodic memory, and describing characteristics of personality. Thus, a system can successfully customize its content and use available resources accordingly by keeping a model for each user [34].

The student model can be observed in the virtual world as an avatar of a real student, the dimensions of the student model match the aspects of the physical student and the properties of the student model represent the characteristics of the real student [35]. Student modeling is one of the key factors affecting automated tutoring systems in teaching decision making [36], as a student model allows students to understand and identify their needs [37]. Student modeling can be defined as the process of collecting relevant information in order to infer and represent the student's current cognitive state in order to be accessible and useful to the adaptation tutoring system [38].

As a result, building an effective student model is a crucial factor in designing an adaptive educational system. To build a student model, it is necessary to consider what information and data should be collected about a student, how it will be updated to keep it up-to-date, and how it will be used for adaptation [39]. In fact, when constructing a student model, the following three questions must be answered: (i) "What are the user's characteristics that we want to model? (ii) 'How are we going to model them?' and (iii) How do we use the model of the user?".

Self [33] identified twenty different uses found in existing ITSs for student models in a review research work. He notes from the analysis of this list that student models' functions could generally be classified into six types:

1. Corrective: to help eradicate bugs in the student's knowledge.
2. Elaborative: to help correct "incomplete" student knowledge.
3. Strategic: to help initiate significant changes in the tutorial strategy other than the tactical decisions of 1 and 2 above.
4. Diagnostic: to help diagnose bugs in the student's knowledge.
5. Predictive: to help determine the student's likely response to tutorial actions.
6. Evaluative: to help assess the student or the ITS.

3.3.1 Student Models Characteristics

3.3.1.1 Modeling Students' Features

The cornerstone of building a student model is the proper selection of the characteristics of the students being performed at their first interaction with the ITS. The aspects of the students being modeled are, according to Gonzalez et al. [40], an initial consideration of the researchers who create an ITS. Domain-dependent and independent characteristics must be taken into account in providing students with effective personalization [35]. Student static features such as email, age, prior knowledge, etc. can also serve as valuable input to the ITS and are determined prior to the learning process [41]. The nature of the static features is to remain unchanged throughout the learning session; however, there are some cases where students may be able to change them through a menu of options available. Furthermore, according to the researchers mentioned above, dynamic features come directly from the interactions of the student with the system and are those that the system constantly updates during learning sessions based on the data collected in the ITS log file.

In view of the above, it is important to determine the characteristics of the dynamic student that constitute the ground for adapting the system to the needs of the individual student. These features may include knowledge and skill levels, errors and misconceptions, styles and preferences of learning, affective and cognitive factors, meta-cognitive factors. Knowledge level refers to a student's prior knowledge of the knowledge domain as well as their current level of knowledge. This is usually measured by tests that must be answered by the student before the learning process. In addition, the system can identify students' misconceptions through these tests along with observing the actions of the student. Learning style refers to individual skills and preferences that affect how learning materials are perceived, gathered and processed by a student [42]. According to Popescu [14], some learners prefer graphic representations, others prefer audio materials and others prefer text representation of learning material, some students prefer to work in groups and others learn better alone. Adapting courses to students' learning preferences has a positive effect on the learning process, resulting in increased efficiency, efficiency and/or satisfaction of learners [43]. The Felder-Silverman learning style (FSLSM) is a proposal for modeling learning styles adopted by many ITSs. FSLSM classifies students in four dimensions: active/reflective, visual/verbal, sensing/intuitive, and sequential/global [44]. The FSLM is then presented and thoroughly discussed. Another method for modeling learning styles is the Myers-Briggs Type Indicator (MBTI) [45], which identifies eight learning styles categories: extrovert, introvert, sensitive, intuitive, thinking, feeling, judging, perceiving.

In traditional classrooms, human tutors monitor and react to the students' emotional state to motivate them and improve their learning process; under the same rationale, an intelligent tutoring system should interpret the students' emotional state and adapt their behavior to their needs, providing a suitable response to those emotions (Lehman et al. 2008). Affective factors are therefore characteristics of the student that should be considered in order to build a model for the student [46]. The affective states may be: happy, sad, angry, interested, frustrated, bored, distracted, concentrated, confused [47]. Rodrigo et al. [48] found some of these emotions, such as boredom or frustration, leading students to behavior off-task. Off-task behavior means that the attention of students is lost and they engage in activities that have nothing to do with the tutoring system and that do not include any learning purpose [49]. Typical examples of off-task behavior include surfing the web, spending time on off-topic readings, talking to order students without learning goals [50]. These behaviors are associated with deep motivational issues [51], thus modeling affective factors can form a basis for motivating students.

Students' cognitive features are important student features that can be held in a student model. These features refer to aspects such as attention, knowledge, learning and understanding skills, memory, perception, concentration, collaborative skills, problem-solving and decision-making skills, analysis of skills, critical thinking. In addition to having cognitive abilities, students need to be able to critically evaluate their knowledge in order to decide what they need to study [52]. Adaptive and/or customized tutoring systems must therefore consider the meta-cognitive abilities of students. Meta-cognition concerns the active monitoring, regulation and orchestration of information processes related to the cognitive objects they carry on [53]. In other words, the notion of meta-cognition deals with the ability of students to be aware of and control their own thinking, such as choosing their learning goals, using prior knowledge or choosing problem-solving strategies intentionally [54]. Some meta-cognitive skills include reflection, self-confidence, self-monitoring, self-regulation, self-explication, self-assessment, and self-management [55].

3.3.1.2 Learning Style Based on Felder-Silverman Model

The Felder-Silverman model (FSLSM) [44] is a model of learning style based on traditional education but also used in computer-assisted education. This chapter will analyze the above-mentioned learning style models in depth along with the FSLSM dimensional characteristics. Through FSLSM's description, it can be clearly stated how the component of student modeling can be improved and improved in any ITS. The ITS can be adapted to the students as such.

For example, a student model that holds information about such characteristics is required to support the adaptation process by incorporating several characteristics of a learning style model to promote adaptivity.

FSLSM exposes a learner's learning style in depth, distinguishing between their dimensional preferences. In addition, FSLSM is based on trends, indicating that sometimes learners with a high preference for certain behaviors can also act differently [56].

FSLSM is very often used in advanced learning technology research related to learning styles. "The Felder Model is the most suitable model for hypermedia courseware," according to Carver et al. [57]. Kuljis and Liu [58] confirmed this by comparing learning style models with the e-learning and web-based tutoring systems application. Graf et al. [56] also suggest FSLSM as the most suitable model of learning style. There are four different dimensions of FSLSM. Each of these dimensions attaches to the student a specific characteristic.

The first dimension distinguishes between an active and reflective method of information processing. Active learners prefer to communicate with their peers and learn to talk about the taught material by working in groups. Reflective learners, on the other hand, prefer working alone.

The second dimension is separated from intuitive learning by sensing. Apprentices interested in a sensing learning style tend to learn facts and practical learning material. They prefer answering questions using known approaches and also tend not to be reluctant with details. In addition, sensing students are more down to earth and use their rationale when they act. They should be more practical than intuitive learners and like to create correlations between the material and reality taught. Intuitive learners, on the other hand, tend to learn abstract teaching concepts like theoretical depictions and their underlying meanings. They are more interested in discerning associations and relationships and tend to have more imagination and original ideas than they are in sensing learners.

The third dimension distinguishes learners between visual learners who can easily recall concepts and who, as such, tend to learn from what they have looked at (e.g., figures, charts and graphs), and verbal learners who can better understand textual representations, whether paper-based or oral.

The fourth dimension characterizes learners based on their preference for learning material to be received and perceived. Sequential learners, having a linear tutoring progress, prefer to learn progressively and incrementally. They present a proneness to progressively make logical steps in understanding the learning material. Global learners, on the other hand, use a holistic process of thinking and learn in great leaps. They prefer to absorb learning material almost randomly, but they suddenly get the whole picture after they have learned enough material. Because the whole picture is important to global learners, they tend to navigate from chapter to chapter through the learning material while sequential learners prefer the learning material to be presented step by step.

Much research has to do with the incorporation of learning styles in adaptive tutoring systems and in education technology in general. Moreover, most tutoring systems that offer user adaptability and focus on learning styles embody only some aspects of these models of learning style and not all of the model's proposed features. The underlying reason is the limitation on specific functions and a specific course structure of most adaptive systems [56]. Therefore, it is important to consider which characteristics of the learning style model are supported by the system when conducting research on learning styles.

There is no need to use all the dimensions to adapt the learning material to students, according to Graf et al. [56]. More specifically, in a specific learning system, not all the characteristic behavior described in the learning style model can be mapped and identified from the behavior. Thus, patterns that indicate specific learning style preferences are tailored to the system features. Therefore, it is important to specify which characteristics can be mapped and identified and which can not be specified when indicating the learning style. Given the characteristics and their relevance to the learning style, a profound estimation of the results of the approach is highlighted and, therefore, a more meaningful application of the information identified.

Comprehensive learning style information is also crucial when identifying relationships between learning styles and student performance in a tutoring system [59] or other student characteristics such as cognitive traits (Graf et al., in press). According to Graf et al. [56], a detailed description of the different characteristics of each dimension and how representative they are is required for that particular dimension of learning style.

3.3.2 *Using a Student Modeling in an ITS*

According to Michaud and McCoy [60], a well-designed tutoring system is actively carrying out two tasks: that of the diagnostic practitioner, discovering the nature and extent of the knowledge of the student, and that of the strategist, planning a response using his learner findings. This is the student model's main role, which is the basis for ITS customization [61]. The system uses the information of a student model to adapt its responses to each individual student, providing dynamically customized instruction, assistance and feedback.

To predict the needs of students and adapt the learning material and process to the learning pace of each individual student, the student model is used for accurate student diagnosis. It is used to produce highly accurate estimates of the level of knowledge and cognitive state of the student to provide them with the most suitable learning material. In addition, an adaptive and/or customized tutoring system can consult the student model to recognize a student's learning style and preferences and make a decision on the learning strategy that is likely to be the most effective for him/her. In addition, an adaptive and/or personalized education system can select suitable learning methods to increase the effectiveness of tutorial interactions and improve learning and motivation by predicting the affective state of the student. In addition, to provide her/him individualized advice and feedback, a student model can be used to identify the strength and weaknesses of the student. In addition, the system can provide the learner with more complicated tasks and proper methods of learning to improve deep learning and help him/her become a better learner by identifying his/her meta-cognitive abilities.

3.4 Platforms for Social e-Learning

In this section, we describe the LMS and CMS platforms selected for review and, in addition, display screenshots of developed prototypes.

A. *Schoology*

Schoology[1] is a cloud based LMS that supporting courses, increased curriculum access and additional content, communication and social networking collaboration (Fig. 3.3).

B. *Moodle*

Moodle[2] offers a single, robust, secure and integrated system for educators, administrators and learners to create personalized learning environment (Fig. 3.4). It has a wide range of innovative and standard features for teaching and learning. In addition, it enables system functionality to be expanded using community-based plugins.

[1] https://www.schoology.com/.

[2] https://moodle.org/.

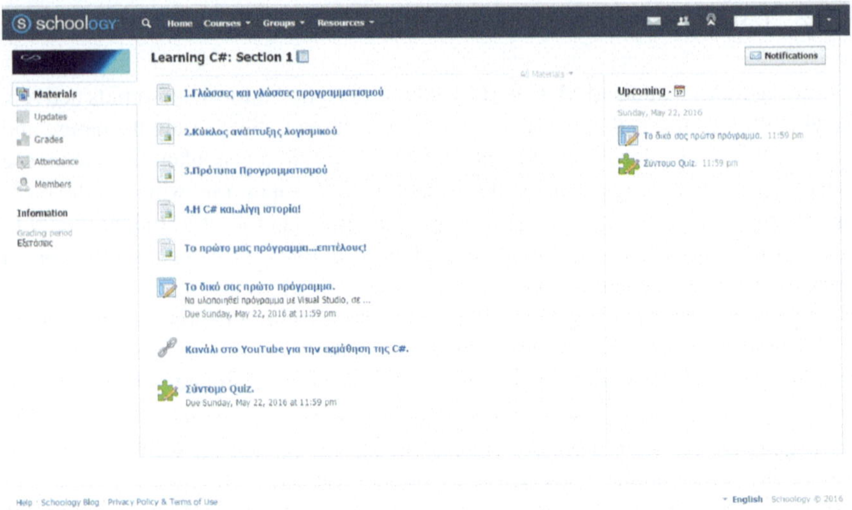

Fig. 3.3 Learning using Schoology

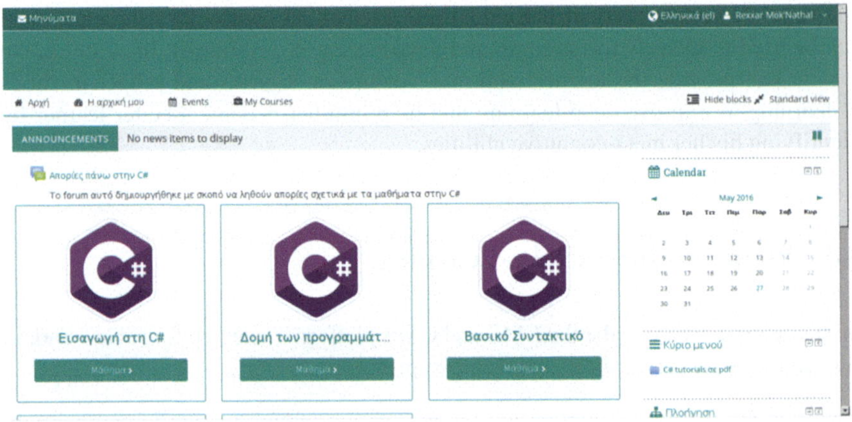

Fig. 3.4 Moodle-based online course

C. *Atutor*

Atutor[3] is a web-based open source LMS for the development and delivery of online courses (Fig. 3.5). Administrators can easily install or update it, customize themes and expand its feature modules easily. Educators can run their courses online quickly and manage the teaching content on the web. Students learn in a social learning environment that is accessible and adaptive.

[3]http://www.atutor.ca/.

Fig. 3.5 Course management in Atutor

D. *Drupal*

Drupal[4] is a content management software that has high standard functionality such as ease of content writing, trustworthy performance, and excellent safety (Fig. 3.6). Free modules are included, which extend and adapt its functionality.

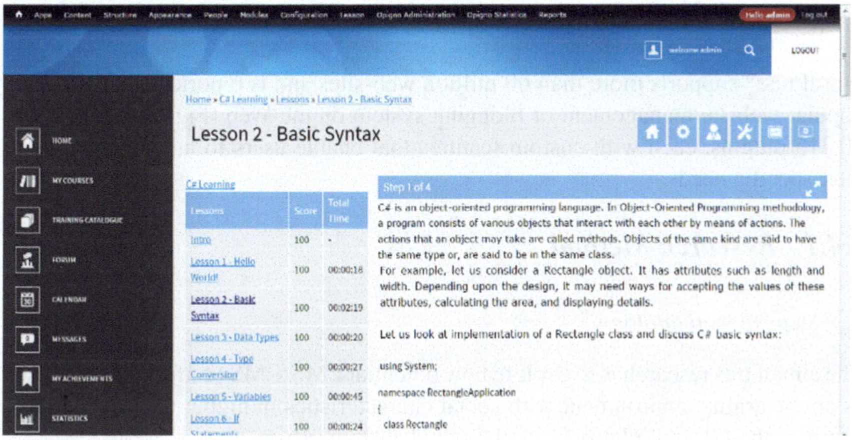

Fig. 3.6 Online course in Drupal

---

[4]https://www.drupal.org/.

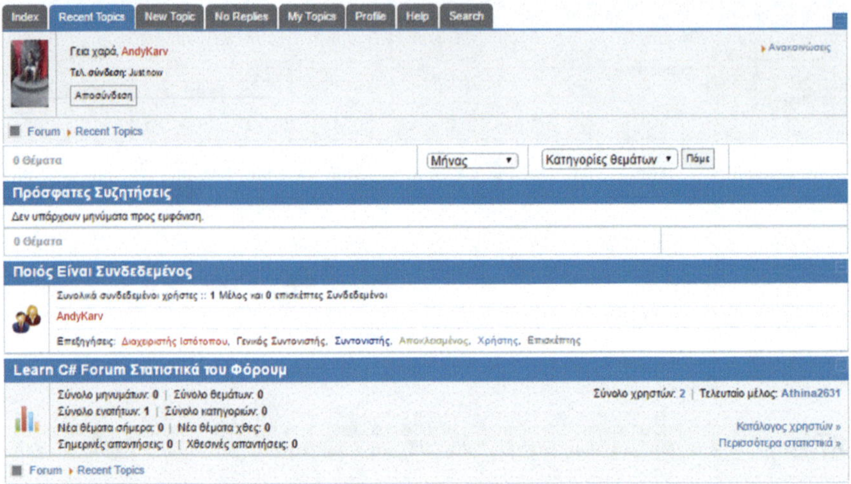

Fig. 3.7 Developing a LMS using Joomla

E. *Joomla*

Joomla[5] is well-known for publishing high-performance online applications (Fig. 3.7). It has an intuitive interface for management of all features and functions. In addition, a variety of free extensions allow users to expand and adapt their functionality to their own aims.

F. *Wordpress*

WordPress[6] supports more than 60 million web-sites and is reported to be the most popular website management or blogging system on the Web (Fig 3.8). It has over 50,316 plugins, each with custom features that enable users to tailor their sites to their specific needs.

3.4.1 Research Method

A. *Steps of methodology*

The aim of this research is to explore how potential LMS/CMS platforms can be used as an e-learning environment with social characteristics in higher education and to compare the selected platforms with several characteristics.

As a first step, we conducted a comprehensive research study focusing on the use of LMS/CMS platforms in the education and learning and LMS/CMS capabilities

[5]https://www.joomla.org/.

[6]https://wordpress.org/.

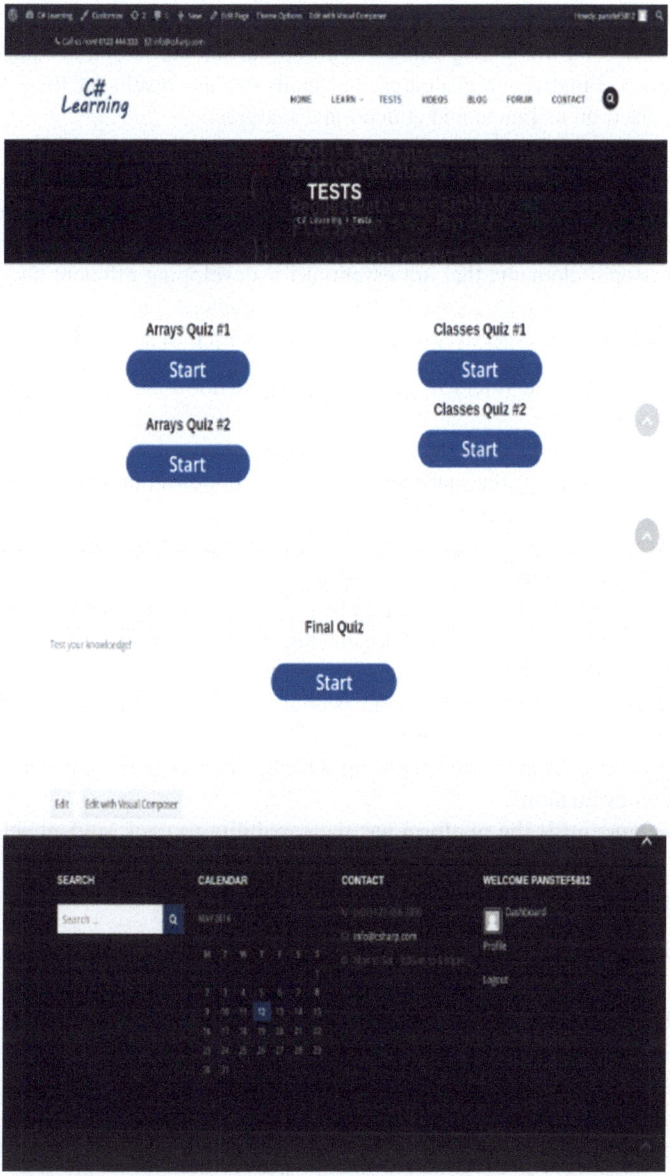

Fig. 3.8 E-learning developed in Wordpress

surveys. We then developed prototypes for selected evaluation LMS/CMS platforms. The course chosen for teaching was C#, a computer science lesson in a huge variety of universities. Finally, we evaluated and analyzed the results of the LMS/CMS platforms based on technical and educational features.

B. *Evaluation criteria*

Several features were chosen, resulting in a more reliable comparison analysis in order to evaluate the LMS/CMS platform. The main technical requirements and basic educational elements that are essential for developing efficient and effective social e-learning systems are covered by these characteristics.

From a technical perspective, the characteristics used in the evaluation are:

- Core functionality: It refers to the type of application the platform mainly has designed to support.
- Open source: If the platform can be freely used, changed and shared.
- Extension modules: If the platform has several modules that can be added on to the site to extend its functionality.
- Customizable: If the platform provides an easy way to customize the site with many widely available configuration options.
- Easy to use: If the platform can be used by users who have no technical experience.
- From an educational perspective, the selected characteristics concern the teaching-learning modules and social interaction, and are:
- Course management: If the instructor is able to create course content, organize it and manage materials distribution.
- Conducting tests: If the platform provides tools for conducting online assessments for student evaluation.
- Tracking student: If the platform has the capability to track student activity and participation.
- Gradebook: If the student grades can be stored and managed by instructor, and reports about learner performance is provided.
- Synchronous and asynchronous interaction: Synchronous learning offers face-to-face interaction, while asynchronous technologies support peer-to-peer one. Examples of synchronous interaction include videoconferencing, webcasts and interactive learning models. On the other hand, asynchronous learning can be carry out even the student is offline and involves coursework delivered via web, e-mail and message boards.
- Social publishing/Communities: If the platform provides the capability of creating communities with common interests or characteristics, where the users can communicate and collaborate effectively through a social interaction, and publishing content by all members, not only by the site's operator (user-generated content).

3.4.2 Comparative Analysis and Discussion

It is a challenging process to develop efficient and effective social e-learning systems by handling the huge number of information exchanged between the participants in the process of learning and their interactions by providing a high quality learning experience. By using LMS/CMS platforms, such systems are easier to develop.

The evaluated systems in this study (LMS/CMS platforms) provide a pleasurable environment for web-based systems development without the requirement for technology and programming. Thus it is done in a simple way without spending a lot of time and effort to deploy an integral web based learning system. In addition, social networking services that support formal and informal learning are included in these platforms.

CMS platforms offer a variety of customization options from a technical point of view, as their main functionality supports Web-based systems of general scope. They have plenty of modules available that can be installed to increase system capabilities. Professional websites can be created using a CMS platform. Instead, LMS platforms lack many extensions and customizable interfaces due to their educational approach.

LMS platforms provide all the modules necessary to promote the training process by default from an educational perspective. Therefore, as required in CMS platforms, instructors do not need to search and install the correct modules. There are a huge number of modules in CMS platforms, however, so that instructors can choose the modules that best suit their needs. In contrast, they offer certain features and limited configuration options thanks to the predefined modules of the LMS platform.

Tables 3.1 and 3.2 illustrates a comparison between LMS/CMS platforms reviewed based on technical and educational criteria.

Table 3.1 Evaluation of LMS platforms

Evaluation aspect	LMS platforms		
Technical	Schoology	Moodle	Atutor
Core functionality	LMS	LMS	LMS
Open source	No	Yes	Yes
Extension modules	It is supported by enterprise version	It provides the capability to install and disable plugins within a single admin interface, freely integrate external applications and content or create your own plugin for custom integrations	There is a central module repository with few modules, integrated feature extensions, or third party add-on software. Developers can create integrated and third party feature modules to extend its functionality
Customizable	Limitations on customization	Customizable site design and functionality though themes and plugins and their easily configuration	Easily create a custom version by importing themes, modifying them, or creating a new one. Except from this, it offers a variety of configurations in modules
Easy to use	It has a simple and easy-to-use interface	It offers a modern and userfriendly interface, easy to navigate on both desktop and mobile devices	It is easy to install, create a system and manage it
Educational			
Course management	It gives the capability to build diverse materials designed to engage students on all levels, set up and organize courses in many different ways	Instructors can selectively release course content and assessments based on specific start and end dates	Instructors can link discussions to specific dates or course events. The system can synchronize course dates defined by the institutional calendar

(continued)

Table 3.1 (continued)

Evaluation aspect	LMS platforms		
Technical	Schoology	Moodle	Atutor
Conducting tests	A basic assessment tool is provided	A tool with huge variety of test options	An integrated system for assessments
Tracking student	It displays student course activity information	Reports for whole student activities	Reports showing student activities
Gradebook	It is provided	It is provided	The system provides test analysis data for individual test items
Synchronous and asynchronous interaction	Instant messaging	Whiteboard. Real-time chats, where chat logs are created and can be shared. Instant messaging	Students can compile selected course content into a downloadable content package for viewing offline. Chatting is supported. It can also be used an accessible instant messaging and white board tool, AComm
Social publishing/Communities	There are groups and discussion features	It provides discussion forums and groupwork	Group functionality is available through the ACollab. Students have personal and public folders which can be shared. Students can create study groups, send e-mail to their groups, use a shared chat space and notice board, and share material privately within the group

Table 3.2 Evaluation of CMS platforms

Evaluation aspect	CMS platforms		
Technical	Drupal	Joomla	Wordpress
Core functionality	CMS	CMS	CMS
Open source	Yes	Yes	Yes
Extension modules	Its functionality can be extended with any one of thousands of add-ons and modules	It is a completely object-oriented software design which allows Joomla! users to write their own extensions and share them with the community. The extensions are divided into plugins, components, and modules	There are over 38,000 plugins in the WordPress Plugins Repository that helps in adding new and enhanced features in a WP site
Customizable	It provides many widely available plugins, themes and other configurations	It offers many customization possibilities for website design and adding functionalities	Due to its widespread popularity, there is a huge variety of customizations, such as plug-ins, themes, etc.
Easy to use	It has a user-friendly interface, so there is no need having programming skills to manage it	It is an intuitive and easy to manage tool without needing technical knowledge	It is easy to install and provides a user-friendly interface
Educational			
Course management	There are appropriate modules	Extension to integrate Joomla and Moodle. LMS plugins	A variety of related plugins
Conducting tests	Through corresponding modules	Many test/quiz plugins	A lot of plugins, provided this capability
Tracking student	No	No	No
Gradebook	Through course and quiz modules	Through test/quiz plugins	Through plugins
Synchronous and asynchronous interaction	Primate messaging, chatrooms	Online Virtual Classroom plugin. Chatting	Live chatting, Messaging
Social publishing/Communities	Forums, Groups	Forums, Groups	Forum, Groups, Commentpress (which allows readers to comment paragraph by paragraph in the margins of a text), Wiki

References

1. Psotka, J., Mutter, S.: Intelligent Tutoring Systems: Lessons Learned. Lawrence Erlbaum Associates, pp. 1–576 (1988)
2. Murray, T.: Authoring intelligent tutoring systems: an analysis of the state of the art. Int. J. Artif. Intell. Educ. **10**, 98–129 (1999)
3. Chrysafiadi, K., Troussas, C., Virvou, M.: A framework for creating automated online adaptive tests using multiple-criteria decision analysis. In: 2018 IEEE International Conference on Systems, Man, and Cybernetics (SMC), pp. 226–231. Miyazaki, Japan (2018)
4. Krouska, A., Troussas, C., Virvou, M.: Social networks as a learning environment: developed applications and comparative analysis. In: 2017 8th International Conference on Information, Intelligence, Systems & Applications (IISA), pp. 1–6. Larnaca (2017)
5. Conde, A., López de Ipiña, K., Larrañaga, M., Garay-Vitoria, N., Irigoyen, E., Ezeiza, A., Rubio, J.: LAGUNTXO: a rule-based intelligent tutoring system oriented to people with intellectual disabilities. In: Proceedings of the 2nd World Summit on the Knowledge Society: Visioning and Engineering the Knowledge Society—A Web Science Perspective, pp. 186–195 (2009)
6. Brusilovsky, P., Millan, E.: User models for adaptive hypermedia and adaptive educational systems. In: Brusilovsky, P., Kobsa, A., Nejdl, W. (eds.) The Adaptive Web. Lecture Notes in Computer Science, 4321, pp. 3–53. Springer, Berlin, Heidelberg (2007)
7. Troussas, C., Chrysafiadi, K., Virvou, M.: Machine learning and fuzzy logic techniques for personalized tutoring of foreign languages. In: Penstein Rosé, C. et al. (eds.) Artificial Intelligence in Education. AIED 2018. Lecture Notes in Computer Science, 10948, pp 358–362. Springer, Cham (2018a)
8. Troussas, C., Krouska, A., Virvou, M., Sougela, E.: Using hierarchical modeling of thinking skills to lead students to higher order cognition and enhance social e-learning. In: 2018 9th International Conference on Information, Intelligence, Systems and Applications (IISA), pp. 1–5. Zakynthos, Greece (2018b)
9. Krouska, A., Troussas, C., Virvou, M., Fragkakis, C.K.: Applying Skinnerian conditioning for shaping skill performance in online tutoring of programming languages. In: 2018 9th International Conference on Information, Intelligence, Systems and Applications (IISA), pp. 1–5. Zakynthos, Greece (2018)
10. Troussas, C., Virvou, M., Alepis, E.: Multifactorial user models for personalized mobile assisted language learning. Frontiers Artif. Intell. Appl. **262**, 275–282 (2014)
11. Troussas, C., Virvou, M., Mesaretzidis, S.: Comparative analysis of algorithms for student characteristics classification using a methodological framework. In: 2015 6th International Conference on Information, Intelligence, Systems and Applications (IISA), pp. 1–5. Corfu (2015)
12. Leung, E., Li, Q.: An experimental study of a personalized learning environment through open-source software tools. IEEE Trans. Educ. **50**(4), 331–337 (2007)
13. Hong, C.M., Chen, C.M., Chang, M.H., Chen, S.C.: Intelligent web-based tutoring system with personalized learning path guidance. In: IEEE International Conference on Advanced Learning Technologies, pp. 512–516 (2007)
14. Popescu, E.: Diagnosing students' learning style in an educational hypermedia system. In: Cognitive and Emotional Processes in Web-Based Education: Integrating Human Factors and Personalization Advances in Web-Based Learning Book Series, pp. 187–208. IGI Global (2009)
15. Alepis, E., Troussas, C.: M-learning programming platform: evaluation in elementary schools. Informatica (Slovenia) **41**(4), 471–478 (2017)
16. Troussas, C., Krouska, A., Virvou, M.: Social interaction through a mobile instant messaging application using geographic location for blended collaborative learning. In: 2017 8th International Conference on Information, Intelligence, Systems & Applications (IISA), pp. 1–5. Larnaca (2017b)
17. Levy, M.: CALL: Context and Conceptualisation, pp. 1–298. Oxford University Press, Oxford (1997)

18. Davies, G., Higgins, J.: Computers, Language and Language Learning. CILT, London (1982)
19. Davies, G., Higgins, J.: Using Computers in Language Learning: A Teacher's Guide. CILT, London (1985)
20. Jones, C., Fortescue, S.: Using Computers in the Language Classroom. Longman, Harlow (1987)
21. Hardisty, D.S., Windeatt, S.: CALL. Oxford University Press, Oxford (1989)
22. Davies, G., Walker, R., Rendall, H., Hewer, S.: Introduction to Computer Assisted Language Learning (CALL): Module 1.4. In Davies, G. (ed.) Information and Communications Technology for Language Teachers (ICT4LT). Thames Valley University, Slough (2011)
23. Warschauer, M.: Computer-assisted language learning: an introduction. In: Fotos, S. (ed.) Multimedia Language Teaching. Logos International, Tokyo (1996)
24. Warschauer, M., Healey, D.: Computers and language learning: an overview. Lang. Teach. **31**, 57–71 (1998)
25. Underwood, J.: Linguistics, Computers and the Language Teacher: A Communicative Approach, pp. 1–109. Newbury House, Rowley, MA (1984)
26. Warschauer, M.: CALL for the 21st Century. In: IATEFL and ESADE Conference, Barcelona, Spain (2000)
27. Bax, S.: CALL—past, present and future. System **31**(1), 13–28 (2003)
28. Decoo, W.: On the mortality of language learning methods. Paper presented as the James L. Barker Lecture at Brigham Young University, Provo, UT (2001)
29. Gimeno-Sanz, A., Davies, G.: CALL software design and implementation. In: Davies, G. (ed.) Information and Communications Technology for Language Teachers (ICT4LT). Thames Valley University, Slough (2010)
30. Egbert, J., Chao, C.C., Hanson-Smith, E.: Introduction: foundations for teaching and learning. In: Egbert, J., Hanson-Smith, E. (eds.) CALL Environments: Research, Practice, and Critical Issues, 2nd edn, pp. 1–14. Alexandria, VA, TESOL (2007)
31. Bangs, P.: Introduction to CALL authoring programs. In: Davies, G. (ed.) Information and Communications Technology for Language Teachers (ICT4LT). Thames Valley University, Slough (2011)
32. Stepp-Greany, J.: Student perceptions on language learning in a technological environment: implications for the new millennium. Lang. Learn. Technol. **6**(1), 165–180 (2002)
33. Self, J.A.: Bypassing the intractable problem of student modelling. In: Frasson, C., Gauthier, G. (eds.) Intelligent Tutoring Systems: At the Crossroads of AI and Education, pp. 107–123 (1990)
34. Kyriacou, D., Scrutable, A.: User modelling infrastructure for enabling life-long user modelling. In: Proceedings of the 5th International Conference on Adaptive Hypermedia and Adaptive Web-Based Systems, pp. 421–425. Hannover, Germany (2008)
35. Yang, G., Kinshuk, K., Graf, S.: A practical student model for a location-aware and context-sensitive Personalized Adaptive Learning System. In: Proceedings of the IEEE Technology for Education Conference, pp. 130–133. Bombay, India (2010)
36. Li, N., Cohen, W.W., Koedinger, K.R., Matsuda, N.A.: Machine learning approach for automatic student model discovery. In: Proceedings of Conference on Educational Data Mining (EDM 2011), pp. 31–40. Eindhoven, The Netherlands (2011)
37. Sucar, L.E., Noguez, J.: Student modeling. In: Pourret, O., Naom, P., Marcot, B. (eds.) Bayesian Networks: A Practical Guide to Applications, pp. 173–185. Wiley, West Sussex (2008)
38. Thomson, D., Mitrovic, A.: Towards a negotiable student model for constraint-based ITSs. In: Proceedings 17th International Conference on Computers in Education, pp. 83–90. Hong Kong (2009)
39. Millán, T., Loboda, T., Pérez-de-la-Cruz, J.L.: Bayesian networks for student model engineering. Comput. Educ. **55**(4), 1663–1683 (2010)
40. Gonzalez, C., Burguillo, J.C., Llamas, M.A.: Qualitative comparison of techniques for student modeling in intelligent tutoring systems. In: Proceedings of the 36th Frontiers in Education Conference, pp. 13–18 (2006)

41. Jeremic, Z., Jovanovic, J., Gasevic, D.: Student modeling and assessment in intelligent tutoring of software patterns. Expert Syst. Appl. **39**(1), 210–222 (2012)
42. Jonassen, D.H., Grabowski, B.L.: Handbook of Individual Differences, Learning and Instruction. Lawrence Erlbaum Associates, Hillsdale, NJ (1993)
43. Popescu, E., Badica, C., Moraret, L.: Accommodating learning styles in an adaptive educational system. Informatica **34**, 451–462 (2010)
44. Felder, R.M., Silverman, L.K.: Learning and teaching styles in engineering education. Eng. Educ. **78**(7), 674–681 (1988)
45. Bishop, C.C., Wheelr, D.: The myers-briggs personality type and its relationship to computer programming. J. Res. Comput. Educ. **26**, 358–371 (1994)
46. Troussas, C., Krouska, A., Virvou, M.: Reinforcement theory combined with a badge system to foster student's performance in e-learning environments. In: 2017 8th International Conference on Information, Intelligence, Systems & Applications (IISA), pp. 1–6. Larnaca (2017a)
47. Balakrishnan, A.: On modeling the affective effect on learning. In: Proceedings of the 5th International Conference on Multi-disciplinary Trends in Artificial Intelligence, pp. 225–235. Hyderabad, India (2011)
48. Rodrigo, M., Baker, R., Maria, L., Sheryl, L., Alexis, M., Sheila, P., Jerry, S., Leima, S., Jessica, S., Sinath, T.: Affect and usage choices in simulation problem solving environments. In: Proceedings of the 13th International Conference on Artificial Intelligence in Education, Marina Del Ray, pp. 145–152. CA, USA (2007)
49. Cetintas, S., Si, L., Xin, Y.P., Hord, C.: Automatic detection of off-task behaviors in intelligent tutoring systems with machine learning techniques. IEEE Trans. Learn. Technol. **3**(3), 228–236 (2010)
50. Baker, R.S., Corbett, A.T., Koedinger, K.R., Wagner, A.Z.: Off-task behavior in the cognitive tutor classroom: when students 'game the system'. In: Proceedings of the SIGCHI Conference on Human Factors in Computing Systems (CHI '04), pp 383–390. Vienna, Austria (2004)
51. Baker, R.S.: Modeling and understanding students' off-task behavior in intelligent tutoring systems. In: Proceedings of the SIGCHI Conference on Human Factors in Computing Systems, pp 1059–1068. San Jose, California, USA (2007)
52. Mitrovic, A., Martin, B.: Evaluating the effects of open student models on learning. In: Proceedings of the 2nd International Conference on Adaptive Hypermedia and Adaptive Web-Based Systems, pp. 296–305 (2006)
53. Flavell, J.H.: Metacognitive aspects of problem solving. In: Resnick, L.B. (ed.) The Nature of Intelligence, pp. 231–236. Hillsdale, Erlbaum (1976)
54. Barak, M.: Motivating self-regulated learning in technology education. Int. J. Technol. Des. Educ. **20**(4), 381–401 (2010)
55. Pena, A., Kayashima, M.: Improving students' meta-cognitive skills within intelligent educational systems: a review. In: Proceedings of the 6th International Conference on Foundations of Augmented Cognition: Directing the Future of Adaptive Systems, pp. 442–451 (2011)
56. Graf, S., Viola, S.R., Leo, T., Kinshuk: In-depth analysis of the Felder-Silverman learning style dimensions. J. Res. Technol. Educ. **40**(1), 79–93 (2007)
57. Carver, C.A., Howard, R.A., Lane, W.D.: Addressing different learning styles through course hypermedia. IEEE Transact. Educ. **42**(1), 33–38 (2009)
58. Kuljis, J., Liu, F.:A comparison of learning style theories on the suitability for elearning. In: Hamza, M.H. (ed.) Proceedings of the IASTED Conference on Web-Technologies, Applications, and Services, pp. 191–197. ACTA Press (2005)
59. Allinson, C., Hayes, J.: The cognitive style index: a measure of intuition analysis for organizational research. J. Manag. Stud. **33**, 119–135 (1996)
60. Michaud, L.N., McCoy, K.F.: Empirical derivation of a sequence of user stereotypes for language learning. User Model. User-Adap. Inter. **14**, 317–350 (2004)
61. Devedzic, V.: Semantic Web and Education, pp. 1–325. Springer, Berlin, Heidelberg (2006)

Chapter 4
Computer-Supported Collaborative Learning: A Novel Framework

Abstract Computer-Supported Collaborative learning deals with the interaction between learners towards working within groups and how collaboration and technology facilitate the sharing and distribution of knowledge among them. This chapter describes the core aspects of computer-supported collaborative learning and presents a machine learning-based hybrid model for win-win collaboration between students which is incorporated in the social networking-based language learning system, called POLYGLOT, which makes adaptive recommendations to students. The recommendation for collaboration concerns two situations. The first situation concerns two students having complementary knowledge, namely student 1 has a high knowledge level on concept A but poor knowledge level on concept B and student 2 has a high knowledge level on concept B but poor knowledge level on concept A. In the second situation, student 1 holds a misconception on A but not B while student 2 holds a misconception on B but not A. This rationale can enhance students in the learning process and ameliorate the degrees of knowledge acquisition and knowledge restitution.

4.1 Computer-Supported Collaborative Learning (CSCL): An Introduction

Computer-supported collaborative learning (CSCL) is a pedagogical approach where learning takes place through the use of a computer or the Internet through social interaction. This type of learning is characterized by participants sharing and building knowledge using technology as their primary means of communication or as a common resource [1]. In online and classroom learning environments, CSCL can be implemented and can occur synchronously or asynchronously.

The study of collaborative learning supported by computers is based on a number of academic disciplines, including education technology, educational psychology, sociology, cognitive psychology and social psychology [2].

CSCL's field draws heavily from a number of learning theories that emphasize that knowledge is the result of learners interacting with each other, knowledge sharing, and knowledge building as a group. Since the field focuses on collaborative activity

C. Troussas and M. Virvou, *Advances in Social Networking-based Learning*, Intelligent Systems Reference Library 181, https://doi.org/10.1007/978-3-030-39130-0_4

and collaborative learning, it takes a lot of constructivist and social cognitive learning theories inherently [3].

4.2 Precursor Theories

The roots of CSCL-related collaborative epistemology can be found in the Social Learning Theory of Vygotsky [4, 5]. The notion of internalization of the theory or the idea that knowledge is developed through one's interaction with one's surrounding culture and society is of particular importance to CSCL [5]. The second important element is what Vygotsky [5] called the Proximal Development Zone. This refers to a range of tasks that can be too difficult for a learner to master on their own, but is made possible with the help of a more skilled person or teacher. These ideas feed into a notion that is central to CSCL, namely by interacting with others, knowledge building is achieved.

Cooperative learning, although different from collaborative learning in some ways, also contributes to team success in CSCL environments. The distinction can be stated as: co-operative learning focuses on the effects of group interaction on individual learning, whereas collaborative learning is more concerned with the cognitive processes in the analysis group unit such as shared meaning making and the common problem space. Positive interdependence, individual accountability, promoting interaction, social skills, and group processing are the five elements identified by Johnson et al. [6] for effective cooperative groups. Comprehension of what encourages successful cooperation is essential for CSCL research because of the inherent relationship between cooperation and collaboration.

Scardamalia and Bereiter [7] wrote seminal articles in the early 1990s leading to the development of key CSCL concepts, namely knowledge-building communities and discourse on knowledge-building, intentional learning, and expert processes. Their work resulted in an early collaboration-enabling technology known as the Computer Supported Intentional Learning Environment (CSILE). A characteristic of CSCL are the fact that the theories have been integrated with CSCL technology design, deployment, and study. Later on, CSILE became the Knowledge Forum, the most widely used CSCL technology in the world.

Other theories of learning that provide a basis for CSCL include distributed cognition, problem-based learning, group cognition, cognitive learning, and situated learning. Each of these learning theories focuses on the social aspect of learning and building knowledge and recognizes that learning and building knowledge involves interpersonal activities including conversation, discussion and negotiation [3].

4.3 Collaboration Theory and Group Cognition

The extent to which computer technology could enhance the collaborative learning process has been explored by researchers over the past two decades. While researchers generally relied on learning theories developed without considering computer support, some suggested that the field needs to have a theory tailored and refined to the unique challenges faced by those trying to understand the complex interplay between technology and collaborative learning [8].

The theory of collaboration, suggested by Stahl [9], as a system of analysis for CSCL, postulates that knowledge is built in social interactions, such as discourse. The theory suggests that learning is not about accepting fixed facts, but is the dynamic, ongoing, and evolving outcome of complex interactions occurring primarily within people's communities. It also emphasizes that collaborative learning is a meaning-building process and that meaning creation occurs most often and can be observed in the analytical group unit. The goal of collaboration theory is to develop an understanding of how meaning is constructed, preserved and re-learned collaboratively in group interaction through language and artifact media. There are four key themes in collaborative theory: collaborative knowledge building (which is viewed as a more concrete term than 'learning'); group and personal perspectives intertwining to create group understanding; artifact mediation (or the use of resources that learners can share or imprint meaning on); and interaction analysis using examples captured that can be analyzed as evidence that knowledge acquisition took place [8].

Collaboration theory suggests that CSCL-supported technology should provide new types of media that foster the building of collaborative knowledge; facilitate the comparison of knowledge based on different types and sizes of groups; and assist collaborative groups in negotiating the knowledge they build. In addition, these technologies and designs should strive in the communication process to remove the teacher as the bottleneck. In other words, the teacher should not have to act as the channel for communication between students or as the channel through which information is given. Finally, collaborative theory-influenced technologies aim to increase the quantity and quality of moments of learning through computer-simulated situations [8].

With the research on group cognition, Stahl [9] extended his proposals on collaboration theory over the next decade. Stahl et al. [1] provided a number of case studies of collaborative technology prototypes, as well as an in-depth sample interaction analysis and several essays on theoretical issues related to the re-conceptualization of cognition in the analysis unit of the small group.

4.4 Strategies

CSCL is currently used in teaching plans in both traditional and online classrooms from primary to post-graduate schools. Like any other educational activity, it has its own prescribed practices and strategies to be used effectively by educators. Because its use is so widespread, the use of CSCL involves countless scenarios, but there are several common strategies that provide a basis for group cognition.

Collaborative writing is one of the most common approaches to CSCL. Although the final product may be anything from a research paper, an online encyclopedia entry, or a short story, the joint planning and writing process encourages students to express their ideas and develop a group understanding of the topic [10]. Tools such as blogs, interactive whiteboards, and custom spaces combining free writing with communication tools can be used to share work, create ideas, and write synchronously [11].

Technology-mediated discourse refers to discussions, discussions, and other techniques of social learning that involve examining a theme using technology. Wikis, for example, are a way for learners to engage in discussion, but other common tools include mind maps, survey systems, and simple message boards. Like collaborative writing, technology-mediated discourse enables participants to engage in conversations and build knowledge together, which can be separated by time and distance [12].

Group exploration refers to the shared discovery between two or more people of a place, activity, environment or topic. Students are exploring in an online environment, using technology to better understand a physical area, or reflecting together through the Internet on their experiences. For this type of learning, virtual worlds as well as synchronous communication tools can be used [13].

Problem-based learning is a well-known educational activity that is well suited to CSCL due to the social implications of problem solving. Complex issues call for rich group interplay that fosters collaboration and moves towards a clear goal [14].

Project-based learning stimulates the development of team roles and the setting of goals. Also, essential for any project is the need for collaboration and encourages team members to build together experience and knowledge. While there are many advantages to using specifically developed software to support collaborative learning or project-based learning in a specific domain, any file sharing or communication tools can be used to facilitate CSCL in problem-or project-based environments [15].

When using Web 2.0 applications (wikis, blogs, RSS feed, collaborative writing, video sharing, social networks, etc.) for computer-supported collaborative learning, specific strategies should be used to implement them, particularly with regard to the following [16]:

- adoption by instructors and learners;
- usability and quality in utilization issues;
- technology maintenance;
- pedagogy and teaching design;
- social interaction between learners;
- privacy issues;
- information/system security of information systems.

4.5 Instructor Roles in CSCL

Although CSCL focuses on people working with their peers, teachers still have a vital role to play in facilitating learning. The instructor must, of course, introduce the CSCL activity in a thoughtful manner that contributes to an overarching course design plan. The design should clearly define the activity's learning outcomes and evaluations. To ensure that learners are aware of these goals and are eventually met, it is necessary to properly manage both resources and expectations in order to avoid overloading learners. Once the activity has started, to facilitate learning, the teacher is charged with kick-starting and monitoring discussion. S/he also needs to be able to mitigate the class's technical problems. Finally, the instructor has to engage in evaluation, in whatever form the design calls for, to ensure that all students have achieved goals.

Any CSCL strategy can lose its effectiveness without the proper structure. It is the teacher's responsibility to make students aware of their goals, how they should interact, potential technological concerns, and the timing of the exercise. This framework should enhance learners' experience by encouraging collaboration and creating knowledge building opportunities. Another important consideration is affordance for educators implementing online learning environments. Students with online communication already comfortable often choose to interact casually. Mediators should pay particular attention to informing online students of their expectations for formality. While sometimes students have reference frames for online communication, they often do not have all the skills needed to solve problems on their own. Ideally, teachers provide what is known as "scaffolding," a knowledge platform on which they can build. A unique advantage of CSCL is that students can use technology to build learning foundations with their peers, given proper teacher facilitation. This allows instructors to assess the difficulty of the presented tasks and to make informed decisions about the extent of the necessary scaffolding [14].

4.6 Effects

The possibility of intellectual partnerships with peers as well as advanced information technology has changed the criteria for what is considered to be the effects of technology, according to Lu et al. [14]. Instead of focusing solely on the quantity and quality of learning outcomes, we need to distinguish between two kinds of effects: "effects with a tool and/or collaborating peers, and their effects." He used the term "effects with" to describe the changes that occur while engaging with peers or with a computer tool. The changed quality of problem-solving in a team can be an example. And it means the word "effects of" lasting changes that occur when computer-enhanced collaboration teaches students to ask more accurate and explicit questions even when they don't use that system.

4.7 Applications of CSCL

It has a number of implications for instructional designers, developers, and teachers.

- First, it revealed what technological features or functions were particularly important and useful to students in writing, and how a CSCL system could be adapted for use in different subject areas, with specific implications to be considered when designing CSCL tools for instructional designers or developers.
- Second, this study also suggested a teacher's important role in scaffolding design, collaborative learning process scaffolding, and making CSCL a success. It is important for CSCL to design a meaningful, real-world task to engage students in authentic knowledge-building learning activities.
- Third, co-operative work in the classroom, using "one-to-one" tool-based technology devices where the teacher has a classroom management program, enables not only teamwork to be enhanced where each member takes responsibility for the group, but also personalized and individualized instruction, adapting to the students' rhythms and achieving the targets set in the classroom.

While CSCL holds promise to enhance education, successful implementation is not without barriers or challenges. Obviously, sufficient access to computer technology is required for students or participants. Although computer access has improved over the past 15 to 20 years, teacher attitudes to technology and sufficient access to internet-connected computers remain barriers to wider use of CSCL pedagogy.

In addition, instructors find that the time required to monitor student discourse and review, comment, and grade student products may be more demanding than what is required for traditional face-to-face classrooms. The teacher or professor also has to make an instructional decision about the complexity of the presented problem. The problem has to be of sufficient complexity to warrant collaborative work, otherwise teamwork is unnecessary. There is also a risk of assuming that students are able to work collaboratively instinctively. Although by nature the task may be collaborative, students may still need to learn how to work in a truly cooperative process.

Others noted a concern about the scripting concept as it relates to CSCL. There is a problem with the possibility of over-scribing the CSCL experience and thus creating "fake collaboration". Such over-written collaboration may not trigger the social, cognitive, and emotional mechanisms needed for true collaborative learning [17].

There is also the concern that the mere availability of the tools of technology may cause problems. Instructors may be tempted to apply technology to a learning activity that can be handled very appropriately without computer intervention or support. They never get to the act of collaboration in the process of students and teachers learning how to use the "user-friendly" technology. As a result, computers are becoming an obstacle rather than a supporter of collaboration [18].

4.8 CSCL for Foreign Language Acquisition

Computer-supported collaborative learning (CSCL) can be traced back to the 1990s as an instructional strategy for second language acquisition. The internet grew rapidly during that time, which was one of the key factors facilitating the process. At the time, the first wikis were still undergoing early development, but the use of other tools such as electronic discussion groups allowed for equal participation amongst peers, particularly benefiting those who would normally not participate otherwise during face-to-face interactions [19].

Global research began to emerge during the establishment of wikis in the 2000s regarding their effectiveness in promoting the acquisition of second language. Some of this research focused on more specific fields such as systemic linguistics, humanistic education, experiential learning, and psycholinguistics. For example, in a class where English was taught as a second language, Chen [20] conducted a study to determine the overall effectiveness of wikis. Another example is Kessler's [21] study in which pre-service, non-native English speaker teachers at a Mexican university were given the task of collaborating on a wiki that served as the final product for one of their courses. In this study, during the course of the task, emphasis was placed on the level of grammatical accuracy achieved by the students.

Other educational tools apart from wikis are being implemented and studied to determine their potential in the acquisition of second language scaffolding due to the continuous development of technology. Blogs, automated writing evaluation systems, and open-source netbooks are among these, according to Warschauer [22]. According to Schmidt [23], the development of other recent online tools has facilitated language acquisition through member-to-member interactions, demonstrating first-hand the impact that technology advancement has had on meeting language learners' varying needs.

4.8.1 *Effectiveness and Perception*

Computer-assisted language learning (CALL) studies have shown that computers provide language learners with material and valuable feedback and can be an effective tool for learning languages both individually and in collaboration. CALL programs provide the potential for language learner-computer interactions [24]. The autonomous language learning and self-assessment of students can also be made widely available on the web. The computer is seen in CSCL not only as a potential language tutor by providing evaluation of the responses of the students, but also as a tool to give language learners the opportunity to learn from the computer and also through collaboration with other language learners. Juan [25] focuses on new models and systems that evaluate student activity efficiently in online education. Their findings indicate that teacher-organized CSCL environments are useful in developing their language skills for students. In addition, CSCL increases the confidence of students and encourages them to maintain active learning, thereby reducing passive reliance on feedback from teachers. It has also been shown to reduce learner anxiety by using CSCL as a tool in the second language learning classroom [26].

In order to measure the effectiveness and perception of CSCL in a language learning classroom, various case studies and projects were conducted. For example, Dooly [27] has shown that language learners have shown that they have increased confidence in using the language and feel more motivated to learn and use the target language. After analyzing the results, Dooly [27] suggests that students have increased awareness of different aspects of the target language during computer-supported collaborative language learning and pay more attention to their own language learning process. Since the participants of this project were trainees of language teachers, she adds that in the future they felt prepared and willing to incorporate online interaction into their own teaching.

4.9 Win-Win Collaboration Module

Furthermore, POLYGLOT assists students concerning the collaboration with peers. As mentioned above, the student model holds information about the students' knowledge level along with the misconceptions that they conduct. To this direction, POLYGLOT supports win-win collaboration. This kind of collaboration is both advantageous and satisfactory to all parties involved. More specifically, both students, who are involved in the collaboration, benefit from the collaboration given that they gain knowledge. This happens as they offer their knowledge and at the same time they receive knowledge. As such, POLYGLOT supports two different kinds of win-win collaboration. The first one concerns the win-win collaboration based on knowledge level and the second one concerns the win-win collaboration based on the nature of misconceptions. Regarding the first kind, POLYGLOT proposes collaboration between two students of whom the student 1 is good at concept A but s/he is not good at concept B and student 2 is good at concept B but s/he is not good at concept A. As an example, if student A achieves a high mark at Chapter 1 of the English language but a low mark in Chapter 2 of the English or French language, POLYGLOT will propose him/her to collaborate with a student who has a low mark in Chapter 1 of the English language but a high mark in Chapter 2 of the English or French language respectively. By the same reasoning, regarding the second kind, POLYGLOT also proposes a collaboration between two students of whom the student 1 makes the error type A but s/he does not make the error type B and student 2 makes the error type B but s/he does not make the error type A. As an example, if student A is prone to make verb tense mistakes but no spelling mistakes, POLYGLOT will propose him/her to collaborate with a student who makes spelling mistakes but no verb tense mistakes. Hence, win-win collaboration can provide a good result for both students involved. Figure 4.1 illustrates how this module works.

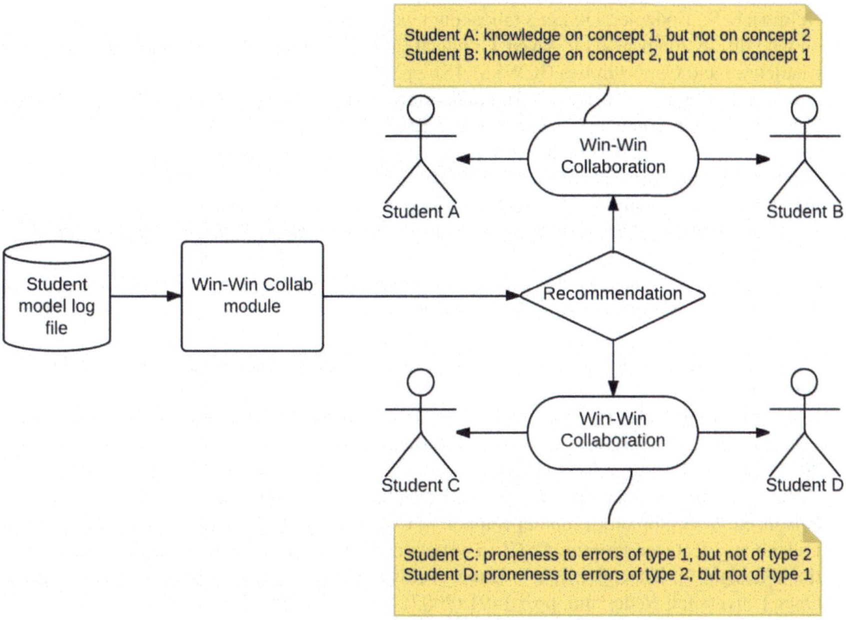

Fig. 4.1 Win-win collaboration module

References

1. Stahl, G., Koschmann, T., Suthers, D.: Computer-supported collaborative learning: an historical perspective. In: Sawyer, R.K. (ed.) Cambridge handbook of the learning sciences, pp. 409–426. Cambridge University Press, Cambridge, UK (2006)
2. Hmelo-Silver, C.E.: Analyzing collaborative learning: multiple approaches to understanding processes and outcomes. In: Proceedings of the 7th international conference on Learning sciences, pp. 1059–1065. USA (2006)
3. Resta, P., Laferriere, T.: Technology in support of collaborative learning. Educ. Psychol. Rev. **19**, 65–83 (2007)
4. Vygotsky, L.: Interaction between learning and development. Read.S Dev. Child. **23**(3), 34–41 (1978)
5. Vygotsky, L.: Mind in society: the development of higher psychological processes. Harvard university press, pp. 1–159 (1980)
6. Johnson, D., Johnson, R., Holubec, E.: Circles of learning: cooperation in the classroom, pp. 95–118. Interaction Book Company, Edina, MN (2002)
7. Scardamalia, M., Bereiter, C.: Computer support for knowledge building communities. J. Learn. Sci. **3**(3), 265–283 (1994)
8. Stahl, G.: Contributions to a theoretical framework for CSCL. In: Stahl, G. (ed.) Computer support for collaborative learning: foundations for a CSCL community—proceedings of CSCL 2002, pp. 62–71 (2002)
9. Stahl, G.: Building collaborative knowing: Elements of a social theory of CSCL. In: Strijbos, J.-W., Kirschner, P., Martens, R. (eds.) What we know about CSCL: and implementing it in higher education, pp. 53–86 (2004)

10. Heimbuch, S., Bodemer, D.: Let's talk about talks: Supporting knowledge exchange processes on wiki discussion pages. In: AAAI Technical Report on Wikipedia, a Social Pedia: Research Challenges and Opportunities (ICWSM-15), pp. 56–61 (2015)
11. Onrubia, J., Engel, A.: Strategies for collaborative writing and phases of knowledge construction in CSCL environments. Comput. Educ. **53**(4), 1256–1265 (2009)
12. Asterhan, C., Schwar, B.: Online moderation of synchronous e-argumentation. Int. J. Comput.-Support. Collab. Learn. **5**(3), 259–282 (2010)
13. Ioannidou, A., Repenning, D., Webb, D., Keyser, L., Luhn, C., Daetwyler, C.: Mr. Vetro: a collective simulation for teaching health science. Int. J. Comput.-Support. Collab. Learn. **5**(2), 141–166 (2010)
14. Lu, J., Lajoie, S., Wiseman, J.: Scaffolding problem-based learning with CSCL tools. Int. J. Comput.-Support. Collab. Learn. **5**(3), 283–298 (2010)
15. Blumenfeld, P., Soloway, E., Marx, R., Krajcik, J., Guzdial, M.M., Palincsar, A.: Motivating project-based learning: sustaining the doing, supporting the learning. Educ. Psychol. **26**(3/4), 369–398 (1991)
16. Bubas, G., Orehovacki, T., Coric, A.: Strategies for implementation of web 2.0 tools in academic education. In: 17th European University Information Systems International Congress (EUNIS 2011), pp. 1–10. Dublin, Ireland (2011)
17. Bannon, L.: Issues in computer supported collaborative learning. In: Proceedings of NATO Advanced Workshop on Computer-Supported Collaborative Learning, pp. 1–13. Italy (1989)
18. Dillenbourg, P.: Over-scripting CSCL: The risks of blending collaborative learning with instructional design. In: Kirschner, P.A.—Three worlds of CSCL. Can we support CSCL?, Heerlen, Open Universiteit Nederland, pp. 61–91 (2002)
19. Ebersbach, A., Glaser, M., Heigl, R.: Wiki: web collaboration, springer science & business media, pp. 1–361 (2008)
20. Chen, Y.: The effect of applying wikis in an english as a foreign language (EFL) class in taiwan. PHD Thesis, University of Central Florida, USA (2009)
21. Kessler, G.: Student-initiated attention to form in wiki-based collaborative writing. Lang. Learn. & Technol. **13**(1), 79–95 (2009)
22. Warschauer, M.: Invited commentary: new tools for teaching writing. Lang. Learn. & Technol. **14**(1), 3–8 (2010)
23. Schmidt, C.: Livemocha and the power of social language learning. J. Educ. Dev. **28** (2010)
24. Chapelle, A.: English language learning and technology, pp. 1–211. John Benjamins Publishing Company, Philadelphia (2003)
25. Juan, A.: Monitoring and assessment in online collaborative environments: emergent computational technologies for e-learning support, pp. 1–324. Information Science Reference, Hershey, PA (2010)
26. Hurd, S.: Anxiety and non-anxiety in a distance language learning environment: the distance factors as a modifying influence. Syst. **35**(4), 487–508 (2007)
27. Dooly, M.: Joining forces: promoting metalinguistic awareness through computer-supported collaborative learning. Lang. Aware. **16**(1), 57–74 (2007)

Chapter 5
Affective Computing and Motivation in Educational Contexts: Data Pre-processing and Ensemble Learning

Abstract Affective computing involves the research area focusing on the implementation of systems and mechanisms that are capable of identifying, understanding and elaborating human affections. It is a field that spans computer science, psychology and cognitive science. Nowadays, affective computing is considered to play a vital role in the field of e-learning since knowledge acquisition can be greatly affected by the changing emotional states of learners. The machines should understand the emotional states of students and adapt their behavior to them, thus providing a tailored response to these emotions. Affective computing systems identify the user's emotional state and react accordingly. In view of the above, this chapter includes a short presentation of the concepts of affective computing tailored to social networking-based learning and learners' affective states. Special mention is made of students' frustration as an emotional state that can either influence the students' learning rates or dropout rates and motivation strategies to overcome problems emerging from negative emotions. This chapter also presents well-known motivational theories as well as pre-processing techniques and ensemble classifiers for sentiment analysis through social networks. Motivation theories concern the support of students to achieve a goal or a certain performance level, leading to goal-directed behaviors. Pre-processing techniques deal with the necessary information to preprocess the reviews in order to find sentiment and make analysis whether it is positive or negative. Finally, Sentiment analysis refers to the use of expert methods (such as natural language processing, text analysis, computational linguistics) to systematically identify, extract, quantify, and study affective states and subjective information.

5.1 Affective Computing

In both traditional and digital learning, emotional states such as frustration may be the reason why students are disappointed or uncomfortable in the learning process [1]. As such, it is important to handle this problem. In traditional learning, where face-to-face learning takes place, when students' affective states are perceived by the instructor, he/she can positively influence them in the tutoring process. In an ITS,

C. Troussas and M. Virvou, *Advances in Social Networking-based Learning*, Intelligent Systems Reference Library 181, https://doi.org/10.1007/978-3-030-39130-0_5

affective states of students should be identified and motivation given to them to tailor the learning content to them.

Picard [2] introduced affective computing for the first time and worked in various fields including gaming, learning, health, and entertainment [3]. Research on affective computing has generated considerable interest in affect detection over the past decade [4]. The affective states used in affective computing are then described and the definition of frustration is presented in detail, which is the affective state taken into consideration for this dissertation.

5.1.1 Affective States

While the representative affective states are related to emotions, feelings and moods, only the emotions are taken into account in the research on affective computing [4]. Traditional theories of emotion involve emotions through expression of the face and body [5, 6]. Darwin [5] first investigated emotions scientifically. Several studies report the correspondence of essential emotions such as fear and rejection with facial and body expressions [4, 5]. Darwin [5] explained six emotions widely acknowledged. Anger, disgust, fear, joy, sadness, and surprise are the six primary emotions [6].

Cognitive psychologists, such as Ortony et al. [7], Roseman [8] and Smith and Ellsworth [9], conducted further emotional research to highlight the close link between emotion and cognition [10]. According to theories of cognitive psychology, such as the theory of evaluation, emotions are determined by people's perception of their experiences and interpretation of an event; the emotions cannot be affected only by experiences and the event. Two people with different evaluations (evaluation of the outcome of the event) or experiences can feel different emotions for the same event in a different environment [11]. This is the pivotal rational of emotional theories of evaluation; they believe that evaluation is seen as the cause of emotion-related cognitive changes [9].

Then, cognitive psychologists explored cognitive approaches to emotions [7–9]. Cognitive psychologists' research aimed at revealing the relationship between variable (circumstances, goal) and emotional labels (joy, fear) [7] and the relationship between evaluation variables and cognitive responses [9]. Theories of cognitive psychology report that emotions are associated with the experience, goal, goal obstruction, goal achievement, etc. of the student. Thereafter, several theories are discussed about cognitive approaches to affective states.

Roseman [8] proposed that five appraisals influence emotions:

- Motivational state: Motivation concerns the rewarding or the avoidance of punishment.
- Situational state: It is related to the presence or absence of the motivational state.
- Probability of achieving the goal.
- Legitimacy: It concerns the deserved outcome of the event.
- Agency: It concerns the outcome and who/what resulted in it.

Based on these appraisals, emotions such as joy, pride, distress, and shame are defined.

Ortony et al. [7] proposed the theory of evaluation, called OCC, and explains emotion as a cognitive assessment of the current situation involving events, agents and objects. Also, the emotions derived from matching a person's preferences or goals with the outcome of the event, namely the reaction to the events, form the emotions according to this theory. Similarly, emotions that occur due to agents and objects are discussed in the OCC theory. The OCC theory defines 22 different emotions, such as "joy," "distress," "hope," "fear" and others, depending on the outcome of the event, the person's interference, and the objects. The cognitive emotions (fear, distress) focus primarily on the goals and outcomes of the student's event. Based on the research of [4, 12–14] reported that the emotion that occurs during the 30–120 min learning sessions is less relevant to the basic emotions. Therefore, the basic emotions may not be relevant to the emotions of the students that occur during the computer interaction. On the other hand, learner-centered emotions such as frustration, boredom, confusion, flow, curiosity and anxiety are more applicable to computer learning environments [4], based on researches by Conati and Maclaren [15], Baker et al. [16], Brawner and Goldberg [17], D'Mello et al. [18], Hussain et al. [19] and Sabourin et al. [20]. In learner-centered affective states, it is important to identify and respond to negative affective states as it may make the student susceptible to quit learning [21]. As far as education research is concerned, there is a controversy as to whether negative affective states such as frustration and confusion are necessary for learning or should be addressed in order to prevent students from quitting it [22]. Research on affective computing satellites that emotional states, such as frustration, can facilitate thinking and learning and are therefore needed during learning [22, 23]. However, Gee [23] further describes that in order to avoid high stress, powerful anger or intense fear, frustration should be kept below a certain level. In addition, frustration causes the disengagement of the student and may eventually lead to attrition [24]. In view of the above, this thesis research focuses on the frustration's negative affective state.

5.2 Frustration as an Affective State

Research on Frustration has been undertaken for over 80 years and concerns a common opposition emotional response. It is associated with anger, annoyance, and disappointment, and arises from the perceived resistance to the fulfillment of the will or goal of an individual, and is likely to increase when a will or goal is denied or blocked. Subsequently, several frustration theories are described.

5.2.1 Rosenweig's Frustration Theory

Rosenzweig [25] defines frustration as the emotion that a person feels in the absence or removal of an ordinarily disposable need or the end-state. For example, if a student interacting with an ITS needs help or just a hint to provide the correct answer in a question but does not receive it, he/she is frustrated because he/she knows it would be easily available in traditional tutoring. The theory reports that frustration can occur as a result of external factors or personal actions, such as the student is unwilling to take the test. In addition, the theory states that "the tolerance of frustration tends to increase with age." Thus, school or college students experience the emotion of frustration more frequently compared to more grown-ups.

5.2.2 Frustration Aggression Hypothesis

Dollard et al. [26] developed the hypothesis of frustration aggression that reports that the frustration experience can always induce some form of aggression. Frustration is defined as a "condition that exists when there is no interference with a goal-response." An example is given to illustrate this definition. Students A are preparing for a test through the study of an ITS' theory. He/she predicts that the goal of achieving high grades is achievable, according to his/her previous interaction with the same course. The rationale for achieving good grades lies in several indicators, such as good test preparation or discontinuing other non-study activities. The power of such indicators can be measured by the likelihood, duration, and strength of high-grade occurrence. A student preparing for exams may be an example. Based on previous experience, he/she predicts that it is possible to achieve the goal of achieving good grades in examinations. His/her interest in achieving the goal is expressed through multiple indicators, such as not playing games, spending less time on social networking activities, and preparing for the examination. The strength of these indicators is measured by the duration, strength and likelihood of the goal-getting good grades occurring. Since force cannot be measured in this example, only consideration is given to the duration of the preparation along with the likelihood of achieving good grades. This example's goal-response, namely the fact that the predicted sequence of the student ends (preparing for the test will lead to good grades) is to achieve good grades. If student A confuses the goal-response with the sequence predicted, he/she will experience frustration. "Aggression is the primary and characteristic reaction to frustration", according to the hypothesis of frustration aggression. Also the hypothesis states that:

- The greater the strength of the goal-response sequence involves, the greater the frustration is; it could affect the strength of the tendency to respond aggressively to frustration.
- The greater the amount of interference with the goal-response is, the greater the tendency to respond aggressively to frustration will be.

- The effect of combined frustration can induce stronger aggressive reaction than individual frustration.

In summary, the severity of frustration is determined either by the amount of the interference or by the strength of the interference or the added effect of several frustrations (cumulative).

5.2.3 Frustration and Goal-Blockage

Morgan et al. [27] defines frustration as "the goal-oriented blocking of behaviour." The main cause of frustration may be environmental, personal, or conflict factors. Environmental factors involve physical impediments that prevent an individual from achieving their goals. Personal factors involve lack of ability, which is necessary to achieve a goal, and the conflict is the inability to achieve a goal because other goals are prioritized. This theory also supports the theory of frustration by Rosenweig; frustration may occur due to external or personal factors in the latter theory. Spector [28] further testifies that frustration can occur when the goal-keeping process is hindered. Frustration occurs when "both goal-oriented interference or goal-oriented activity and goal-maintenance interference" occurs. In other words, a person will experience frustration if any goal or expected outcome is hindered. In addition, a person will be frustrated if he/she continues to maintain his/her goals. The factors affecting the frustration's strength are the importance of the impeded goal, the degree of interference, and the number of interferences that impede the achievement of the goal. Cognitive psychologists Ortony et al. [7], Roseman [8] and Smith and Ellsworth [9] perceive assessment as the reason for cognitive emotions that occur from the perspective and expectation of an event by a person. Cognitive psychologist theories state that emotions are related to the experience, goal, goal obstruction, goal attainment, etc. of the student. As such, when the goal matches the outcome of an event, emotions are revealed.

5.2.4 Frustration and Cause in Computer Users

In addition to the theories of frustration, the following frustration attribute has been studied in the related scientific literature and is not explicitly mentioned in the theories of frustration. Lazar et al. [29] investigated the causes of computer users' frustration. Their research reports that it is directly proportionate to a higher level of frustration that a task is of higher importance for students to spend a high amount of their time. Indeed, if an important goal is not achieved for the preparation the student spent a lot of time, it leads to a higher degree of frustration. Under this rationale, to detect frustration, the time spent attaining a goal is significant.

5.2.4.1 Definition of Frustration as Used in This Book

The main and underlying reason for frustration, as mentioned above, is the impediment of the goal. Identifying the primary reason for frustration is therefore vital. Besides this, there are also more reasons for frustration in the theories mentioned above except for external interference. External factors, such as computer hardware problems or even connectivity issues, will not be a reason for frustration in the field of e-learning. This research takes into account the following reasons for frustration based on Dollard et al. [26], Lazar et al. [29], Morgan et al. [27] and Spector [28] to model the frustration of the students:

- Frustration is the blocking of a behavior directed towards a goal.
- The distance to the goal is a factor that influences frustration.
- Frustration is cumulative in nature.
- Time spent to achieve the goal is a factor that influences frustration.
- Frustration is considered as a negative emotion, because it interferes with a student's desire to attain a goal.

5.3 Motivation Theory

Affective computing involves the detection and responsive action of the affective state of the student. In the previous sections, the way affective states are detected has been described in depth. The response to the affective state of the student is subsequently presented by displaying motivational messages. In order not to experience frustration or goal failure, motivational messages are used to urge the learner to study using the ITS. Motivation theories are described in this section. The theories of motivation are used to motivate the person to work or to continue to work. Psychologists of motivation report that the desire to achieve is the cornerstone of the theories of motivation. By investigating the application of the motivation theory to the outcome of the event (either success or failure) [30], cognitive motivational theories were developed. The motivation theories that dominated the scientific motivation study are briefly presented in this section.

5.3.1 Hull's Drive Theory

The drive theory of the Hull was the first motivation theory and is based on the energy (drive) needed to motivate the individual [31]. At the same time, it coincides with a characteristic of the educational process, namely if the response to a stimulus (action to an event and response to that event-goal-response) ends with a satisfactory outcome, the motivation increases; if it ends with an annoying outcome, the motivation decreases. According to this theory, the habit is the strength needed to increase the motivation, which is reduced as a result of the stimulus response. In other words, a habit is the action a person requires to pursue the goal. The habit can, however, provide the necessary directions for an action, but not the drive. The mathematical relationship for motivational behavior between drive and habit is therefore given below:

- Behavior = Drive * Habit

Behavior is proportionate to Drive and Habit. This is to indicate that only Drive or Habit alone cannot motivate the person. If there is no energy (Drive = 0), the person would not act irrespective of the strength of the habit.

5.3.2 *Lewin's Field Theory*

Kurt Lewin's theory[1] is based on Gestalt psychology[2] in order to interpret the motivational behavior, being known as the field theory. The Gestalt psychology analyzes the behavior as a whole and is not determined by the summation of individual elements. The field theory states that behavior is determined by both the person and the environment involved:

Behavior = f(Person, environment)

The motivational force of a person is associated with three factors:

- the person's intent (need) to complete the task, known as tension (t)
- The magnitude of the goal (G), which satisfies the need and
- The psychological distance of the person from the goal (e).

The mathematical function for the motivational force of a person is: Force = f(t, G)/e. In this function, the psychological distance from the goal is inversely proportionate to the motivation force; namely, if the distance to achieve the goal is reduced (approaching zero), then the motivation to achieve the goal is increased.

5.3.3 *Atkinson's Theory of Achievement Motivation*

Following the same rationale of the aforementioned theories, Atkinson also developed the mathematical function for achieving motivation; however, Atkinson focused

[1] http://www.psychologydiscussion.net/learning/learning-theory/lewins-field-theory-of-learning-education/2525.

[2] http://webspace.ship.edu/cgboer/gestalt.html.

on individual differences in motivation. Atkinson's theory[3] states that the behavior (tendency) to approach an achievement-related goal (T_t) is the product of three factors:

1. the need for achievement or motive for success (M_S),
2. the probability that a person will be successful at the task (P_S) and
3. the incentive for the success (I_S). The mathematical function is: $Ts = M_S * P_S * I_S$

The achievement motive M_S is developed during the early stages of life and shaped by child-reading practices. The probability of success Pi usually defined in terms of the difficulty of the task. The value of Pi ranges from 0 to 1. The third factor, which is the incentive of success I_S, is inversely related to P_I: $I_S = 1 - P_S$.

5.3.4 *Rotter's Social Learning Theory*

Rotter's theory[4] is also based on individual differences in behavior, like the Atkinson's theory. The motivational model by Rotter is based on the general expectancy (E) and reinforcement value (RV), and the relationship of these two factors is:

Behavior = f(E, RV)

Reinforcement value (RV) is a comparative term and is not clearly mentioned in the theory [30]. The expectancy (E) of success depends on one's history of the present situation and similar circumstances. For example, one's expectancy of success in an event depends on the history of success or failure in the same event or the result of similar events. In a situation which requires one's skill, the expectancy increases after success and decreases after failure.

5.3.5 *Attribution Theory*

The Attribution Theory attempts to explain the world and determine the cause of an event or behavior (for example, why people do what they do). When applied to motivation, the theory of attribution considers the expectation of the person and the response from the event. Heider [32] constructed this theory and Weiner [33] later developed this theory. The theory of attribution [33] relates emotional conduct to academic (cognitive) success and failure. The causes of success and failure are analyzed in conjunction with the context of achievement. The person's reaction is related to an event's outcome. As such, if the outcome is successful and frustrated

[3] https://principlesoflearning.wordpress.com/dissertation/chapter-3-literature-review-2/the-human-perspective/achievement-motivation-atkinson-mcclelland-1953/.

[4] http://psych.fullerton.edu/jmearns/rotter.htm.

or sad, a person feels happy if the outcome of the event fails. This is referred to as "independent allocation of results" [33]. The learner's attribution of success or failure is analyzed in three sets of locus, stability, and controllability characteristics.

- Locus refers to the location of the cause, which can be internal or external to the cause of success or failure. Locus determines whether the outcome of an event (success or failure) alters pride and self-esteem. If the learner attributes the success to internal causes such as being well prepared for the examination and doing more homework, then it will lead to pride and motivate the learner to set new objectives. Whereas, if the learner attributes the failure to internal causes, the self-esteem will be diminished. Therefore, the failure of the learner should be attributed to external factors, such as hard testing or language learning difficulties, in order to motivate the learner to focus on future events.
- Stability refers to the future performance of the learner. If the learner attributes success to stable factors such as "low ability," then, given the same environment, the outcome of the future event will be the same. The future success is unlikely if the learner attributes the failure to the stable factors. If the learner attributes the failure to unstable factors like "less effort" and "luck" then the success of the learner will be improved in future events [34].
- Controllability refers to the factors that the learner who has the ability to change them can control. This will lead to self-motivation if the learner failed the task but can control the future outcome by altering it, such as improving math-solving ability, spending more time on homework. On the other hand, this will lead to shame or anger if the student is unable to control a failure at a task.

The Attribution Theory states that the attribution of a person to success or failure contributes to the future activity effort of the person. If the learner attributes the success to factors that are internal, stable and controllable, it will result in pride and motivation. If the learner attributes the failure to the internal, stable, and uncontrollable factors, then the self-esteem, shame, and anger will be reduced. Motivating the failure of the students with messages attributing the failure to external or unstable or controllable factors will therefore help them set a new goal with self-motivation.

5.3.6 Discussion on Motivational Theories

Hull's drive theory and Lewin's field theory both explain what determines motivation by using the same factors: a person's need (Hull's "drive" and Lewin's "tension"), goal object, and directional value (Hull's "habit" and Lewin's "psychological distance"). Subsequently, these factors are not considered either in Atkinson's and Rotter's theories of expectancy-value or in the theory of attribution. Atkinson's motivation for achievement and Rotter's theory of social learning focus on the motivation, success rate, and history of the individual. These theories, however, are addressed to the broader motivation goals and did not offer suggestions to increase performance in the classroom. Therefore, traditional and digital learning are not tailored to them.

Graham et al. [30] conducted a research reviewing the above-mentioned theories and reported that each theory had a lifespan of approximately 20 years and significant contributions to the theories were made over this period. Hull, Lewin and Atkinson's theories were not used after their span of life. Research on the social learning theory of Rotter has also been reduced. Research on the theory of attribution and its application to achievement seems to dominate motivation theory [30]. Graham [35] also reviewed the papers on the theories of motivation. This study reports that (a) in that decade there were 66 published studies and the primary conceptual framework was the theory of attribution and (b) "Attribution Theory has been proven to be a useful conceptual framework for studying motivation in educational contexts".

More recently (1990 onwards), the theory of motivation for its applications has been investigated. For instance, the theory of self-determination [3, 36, 37] is an application of organizational motivation theory. The theory of self-determination discusses the relevance in organizational behavior of work motivation. The expectation-value motivation theory of achievement [38] relates to the motivation of the child's expectation of success, ability, and subject task. The Attribution Theory is, of course, the theory that fits perfectly in the field of education [39, 40]. In affective computing, in particular in ITS, Attribution Theory is also used to address the affective states of the students [13, 41]. In this research, therefore, the theory of attribution was selected to create motivational messages and address the affective states.

5.4 Responding to Frustration

In this section, the various approaches used in computer-based learning environments to respond to frustration are presented. Klein et al. [42] outlined the strategies for responding to affective states of students. Based on previous research work on active listening [43, 44], these strategies are developed. The guidelines to respond to affective states listed in Klein et al. [42] are as follows:

- The system should provide an option for the student to receive feedback based on their affective state. This is to show the student that their emotions are being actively listened to by the system. Active listening to the emotions of students has shown that their emotions have changed [43].
- Feedback from the students should be promptly requested whenever the student is frustrated. The request for feedback will be ineffective if the student is not frustrated. The students should have a list of options to choose from to report the affective states. This will provide the student with the option to react to what he/she is experiencing emotion.
- The system should provide feedback messages showing empathy, which should make the student feel that he/she is not alone in that affective state. The messages should also convey to the student the validity of the emotion he/she is experiencing. For example, the student should not feel that only he/she has had incorrect answers to the system question or that he/she has missed the goal.

Other approaches to responding to affective states include displaying agent messages [45, 46]. The agents are designed to demonstrate empathy and encourage students to continue learning. In addition, the positive messages to address the emotion of the students helped them improve their performance [47]. This dissertation is based on the research of Dweck [48, 49] on feedback messages to praise the student's effort rather than the intelligence of the student in order to create motivational messages. A nonverbal IQ (Intelligence Quotient) test was performed on students in these researches and provided one of three forms of feedback messages. One-third of students have been praised for their intelligence, one-third of students have been praised for their efforts and the remaining students have not been praised for effort or intelligence. The students were given second set of problems that are difficult compared to the first set of problems after providing the feedback message. The students were interviewed later to learn about their intelligence views. The result shows that the students praised for intelligence believe the intelligence is fixed and cannot be improved. The students, who have been praised for their effort, believe that more effort can enhance intelligence. The students, who have been praised for their efforts, also believe that failure means low effort and showed more pleasure in solving difficult problems. Research on feedback messages by the Dweck is a seminal work in the research area of guidelines for creating feedback messages and has been applied to a wide range of educational research (examples include motivating school students [50] and responding to the affective states of students in computer-based learning [12, 16]. All the above approaches for responding to frustration are adapted in this research. The content is based on the theory of attribution in our motivational messages [33]. The option for students to reflect their feedback is provided on the basis of Klein et al. [42] guidelines; feedback is requested after detection of frustration and feedback messages to show empathy for the affective state of the students. Using the recommendation in Hone [46] and Prendinger and Ishizuka [45], the motivational messages are displayed using an agent that provides empathy in the messages displayed. The motivational messages are constructed on the basis of Dweck's research [49] to praise the effort of the students and not (only) their intelligence.

5.5 Pre-processing Techniques and Ensemble Classifiers for Sentiment Analysis Through Social Networks

5.5.1 Methodology

The objective of the current research is to address the following research questions:

1. What role do data preprocessing techniques play in classification issues?
2. Which stand-alone classifier is the most accurate result?
3. Is the algorithm ensemble outperforming the stand-alone ones?
4. Does the use of multiple data sets show different performance indicators on different domains?

To answer these questions, different data pre-processing options and a range of well-known classifiers were used to conduct a set of experiments. All experiments were conducted in three Twitter datasets: the Obama-McCain Debate (OMD) dataset, the HCR dataset, and the Stanford Twitter Sentiment Gold Standard (STS-Gold) dataset. First, we use various preprocessing options to examine the performance of several well-known learning-based classification algorithms. For tweets, a weighting scheme, a stemming library and a removal list of stop-words were applied. It was then selected to compare three different tokenization settings: unigram, bigram, and 1-to-3-gram. For each of these options, extraction methods of features were applied to estimate whether eliminating poorly characterizing attributes can be useful for better classification accuracy. The methods used were: one based on gaining information and the other based on the classification of Random Forest. The results of these experiments were used to perform the proper preprocessing options for subsequent classifier comparative analysis. Four representative and state-of-the-art machine learning algorithms were selected for the second research issue, covering different classification approaches. Naïve Bayes (NB), Support Vector Machine (SVM), K-Nearest Neighbors (KNN) and Decision Tree (C4.5) are the classifiers. Two commonly used methods were implemented for the validation phase: split percentage and cross validation k-fold. We run the ensemble algorithms, namely Bagging, AdaboostM1, Stacking and Vote, using the four base classifiers mentioned above, towards the combination of sentiment analysis algorithms. Compared to the results of the previous analysis of stand-alone classifiers, these ensembles were compared.

The confusion matrix, one of the most popular tools, was the evaluation model used in this research.[5] Its focus is on a model's predictive capacity rather than how quickly it takes the model to perform classification, scalability, etc. The matrix of confusion is represented by a matrix representing the instances in a predicted class in each row, whereas each column is represented in an actual class. The confusion matrix derives from various measures, such as error rate, accuracy, specificity, sensitivity, and accuracy, and several advanced measures, such as ROC and Precision-Recall. One of the advantages of using this performance evaluation tool is that it can be found easily if the model confuses two classes (i.e. it usually mislabels each other). The matrix also shows the classifier's accuracy as the percentage of properly classified patterns in a class divided by that class's total number of patterns. The classifier's overall (average) accuracy is also assessed using the confusion matrix.[6]

The preprocessing settings and the learning-based algorithms were executed using Weka data mining package.[7] The outcomes of the implementation have been tabulated. Afterwards, a descriptive analysis has been conducted to answer to research issues.

[5]http://rali.iro.umontreal.ca/rali/sites/default/files/publis/SokolovaLapalme-JIPM09.pdf.

[6]http://aimotion.blogspot.gr/2010/08/tools-for-machine-learning-performance.html.

[7]http://www.cs.waikato.ac.nz/ml/weka/index.html.

Table 5.1 Statistics of the three Twitter datasets used

Dataset	Tweets	Positive	Negative
Obama-McCain Debate (OMD)	1904	709	1195
Health Care Reform (HCR)	1922	541	1381
Stanford Twitter Sentiment Gold Standard (STS-Gold)	2034	632	1402

5.5.2 *Twitter Datasets*

In the current experiments, three well-known Twitter datasets have been used; the Obama-McCain Debate (OMD) dataset [51], the Health Care Reform (HCR) dataset [52] and the Stanford Twitter Sentiment Gold Standard (STS-Gold) dataset [53]. The reason they were chosen is because they are available free of charge on the web and were created with a significant number of tweets by reputable universities for academic scope. In addition, they were used in various investigations [54]. Thus, for our experiments, these datasets are considered reliable. The data sets statistics are displayed in Table 5.1.

Obama-McCain Debate (OMD). The Obama-McCain Debate (OMD) dataset was built from 3238 tweets crawled in September 2008 during the first U.S. presidential television debate [51]. Using Amazon Mechanical Turk, sentiment labels were acquired. The set used in this paper was 709 positive and 1195 negative, which was agreed upon by two-thirds of the voters.

Health Care Reform (HCR). The Health Care Reform (HCR) dataset was built in March 2010 by hashtag #hcr tweets [16]. The authors manually annotated a set of 2516 tweets with five labels: positive, negative, neutral, irrelevant, uncertain. A subset of 1922 tweets was considered for this research, excluding irrelevant, uncertain and neutral tweets labeled. Therefore, 541 positive and 1381 negative tweets were included in the final dataset.

Stanford Sentiment Gold Standard (STS-Gold). The STS-Gold dataset was created by selecting tweets from Stanford Twitter Sentiment Corpus 10 and includes independent tweet and entity sentiment labels to support tweet-based and entity-based Twitter sentiment analysis models [53]. The set of 2034 tweets has been used in current experiments with 632 positive and 1402 negative ones.

5.5.3 *Evaluation of Data Preprocessing Techniques*

Preprocessing is a necessary step in preparing data for classification of feelings. We use the StringToWordVector filter to perform preprocessing in WEKA (Table 5.2). This filter allows configurations as follows:

Table 5.2 Preprocessing techniques applied

Preprocessing technique	Applied option
Weighting scheme	TF-IDF
Stemming	Snowball stemmer
Stop-words removal	Rainbow list
Tokenization	1. Unigram 2. Bigram 3. 1-to-3-gram
Feature selection	1. All 2. InfoGainAttributeEval/Ranker—IG >0 3. ClassifierAttributeEval-RadomForest/Ranker—top 70%

- **TF-IDF weighting scheme**: It is a standard approach to building vectors. TF-IDF stands for the "term frequency-inverse frequency document frequency" and is a numerical statistic that reflects the importance of a word in a corpus document.
- **Stemming**: Stemming algorithms work, according to some grammatical rules, by removing the word suffix. We use the Snowball stemmer library in this study, which is the most popular and standard approach.
- **Stop-words removal**: It is a technique that eliminates words of frequent use that are meaningless and useless to classify text. This reduces the size of the corpus without losing significant information. For our experiments, the Rainbow list is used.
- **Tokenization**: This setting divides the documents into words/terms, creating a bag-of-words word vector. We propose to compare word unigram, bigram and 1-to-3-gram with NGramTokenizer.

The preprocessing above generates an enormous number of attributes, many of which are not relevant to classification. Therefore, the following operation is applied:

- **Feature selection**: It is a process that reduces the number of attributes into a better subset that can bring the highest precision. The advantages of executing this data option are limiting overfitting, improving accuracy, and reducing training time. It is possible to classify feature selection methods as filters and wrappers. Filters are based on statistical tests such as Infogain, Chisquare and CFS, whereas Wrappers use a learning algorithm to report the optimal feature sub-set. WEKA provides the AttributeSelection filter to select an attribute evaluation method and a search strategy for this task. We are examining three options in this paper:

  a. No filter applied. We use all attributes created by StringToWordVector filter.
  b. InfoGainAttributeEval, which evaluates the worth of an attribute by measuring the information gain with respect to the class and we set the Ranker search method to select attributes with IG >0, and
  c. ClassifierAttributeEval, which evaluates the worth of an attribute by using a user-specified classifier. We choose Random Forest as classifier and set the Ranker search method to select the top 70% attributes.

Table 5.3 Number of attributes created by attribute selection option

Dataset	Attribute selection	N-gram		
		Unigram	Bigram	1–3 g
OMD	All	2150	7400	2430
	IG >0	264	519	1074
	Top 70%	1500	5180	1680
HCR	All	3000	1835	2945
	IG >0	281	645	1280
	Top 70%	2100	1280	2060
STS-Gold	All	2990	8420	2115
	IG >0	252	354	720
	Top 70%	2090	5890	1480

We conducted a variety of experiments testing the options that would return more accurate results to specify the optimal settings of the preprocessing techniques and classifiers. The preprocessing methods selected, described above, were applied to the three Twitter datasets and the Naïve Bayes Multinomial (NBM), nu-SVM type, KNN with k = 19 and default C4.5 settings were selected for classification. The 10-fold cross validation method has been used for the validation phase.

A different number of attributes were created based on which the classification was performed, based on n-gram and attribute selection options. Table 5.3 shows the numbers. We note that selecting attributes with information gain above zero will significantly decrease the resulting number of attributes. For the second extraction feature, where Radom Forest algorithm evaluates the attributes, we select approximately 70% of the attributes ranked as worthy. An expected advantage of selecting attributes is that algorithms train more quickly.

Table 5.4 shows classifier performance depending on the methods used for preprocessing. The behavior is not uniform when it comes to representations of the dataset. There is no representation that systematically delivers better results compared to the others. In general, with a close competition with unigram, 1-to-3-grams perform better than the other representations.

Our evaluation results show that the operation of selection of attributes improves classification performance over the selection of all attributes. This results from removing redundant and irrelevant attributes from the datasets that may be misleading to modeling algorithms, resulting in overfitting. In Table 5.5, we note that when applying the selection of attributes on the basis of information gain, significant accuracy rates are obtained. Note that the percentage of correctly classified instances is increased by more than 7 points in the case of NB. In addition, as an attribute selection classifier, the Random Forest algorithm enhances classification accuracy compared to all attributes. Lastly, there is no improvement in the performance of algorithms in some experiment settings by applying a selection filter attribute, but this is of insignificant value because the divergence is too low.

Table 5.4 Classifiers' confusion matrices related with preprocessing techniques

N-gram	Attribute Selection	Measures	Dataset											
			OMD dataset				HCR dataset				STS-Gold dataset			
			Classifier				Classifiers				Classifiers			
			NB	SVM	KNN	C4.5	NB	SVM	KNN	C4.5	NB	SVM	KNN	C4.5
Unigram	All	Acc. (%)	79.46	81.93	73.32	75.26	77.59	**83.15**	72.02	73.63	82.10	83.63	68.93	74.29
		Pr.	0.792	0.831	0.785	0.750	0.763	0.852	0.799	0.712	0.817	0.850	0.475	0.731
		R.	0.795	0.819	0.733	0.753	0.776	0.832	0.720	0.736	0.821	0.836	0.689	0.743
		FM.	0.793	0.809	0.689	0.742	0.763	0.808	0.605	0.714	0.814	0.821	0.562	0.711
	IG>0	Acc. (%)	86.61	**83.61**	**74.21**	**75.89**	85.34	79.04	**72.85**	**74.57**	88.35	**83.92**	**73.40**	**74.29**
		Pr.	0.868	0.860	0.735	0.762	0.853	0.839	0.815	0.731	0.882	0.852	0.776	0.734
		R.	0.866	0.836	0.691	0.759	0.851	0.799	0.771	0.751	0.883	0.839	0.734	0.743
		FM.	0.863	0.825	0.568	0.744	0.847	0.782	0.693	0.694	0.881	0.825	0.663	0.707
	Top 70%	Acc. (%)	**88.08**	83.25	73.58	74.79	**88.25**	82.53	72.02	73.27	**91.40**	83.68	68.93	74.04
		Pr.	0.880	0.852	0.748	0.748	0.884	0.853	0.799	0.707	0.914	0.854	0.475	0.732
		R.	0.881	0.832	0.736	0.748	0.882	0.825	0.720	0.733	0.914	0.837	0.689	0.740
		FM.	0.880	0.822	0.709	0.733	0.876	0.797	0.605	0.709	0.912	0.820	0.562	0.701
Bigrams	All	Acc. (%)	75.26	85.82	62.76	**69.22**	82.16	78.00	71.81	71.66	70.30	**85.05**	68.93	70.50
		Pr.	0.736	0.882	0.394	0.728	0.822	0.815	0.516	0.659	0.680	0.877	0.475	0.702
		R.	0.723	0.858	0.628	0.692	0.822	0.780	0.718	0.717	0.703	0.851	0.689	0.705
		FM.	0.692	0.849	0.484	0.632	0.822	0.728	0.600	0.635	0.624	0.834	0.562	0.616
	IG>0	Acc. (%)	**89.02**	82.88	**63.39**	68.28	**89.13**	**78.36**	71.81	71.92	**86.77**	80.63	**68.98**	70.21
		Pr.	0.908	0.848	0.415	0.731	0.895	0.834	0.516	0.672	0.874	0.836	0.683	0.697

(continued)

Table 5.4 (continued)

N-gram	Attribute Selection	Measures	Dataset											
			OMD dataset				HCR dataset				STS-Gold dataset			
			Classifier				Classifiers				Classifiers			
			NB	SVM	KNN	C4.5	NB	SVM	KNN	C4.5	NB	SVM	KNN	C4.5
		R.	0.898	0.829	0.632	0.691	0.891	0.784	0.718	0.719	0.868	0.806	0.690	0.702
		FM.	0.890	0.822	0.488	0.625	0.885	0.730	0.600	0.612	0.860	0.778	0.565	0.608
	Top 70%	Acc. (%)	83.04	**87.08**	62.76	69.01	87.52	78.16	**71.82**	**72.07**	80.53	84.32	68.93	**70.55**
		Pr.	0.853	0.892	0.394	0.747	0.873	0.827	0.516	0.682	0.837	0.872	0.475	0.705
		R.	0.830	0.871	0.628	0.690	0.875	0.782	0.718	0.721	0.805	0.843	0.689	0.706
		FM.	0.818	0.863	0.484	0.621	0.871	0.728	0.600	0.619	0.776	0.825	0.562	0.616
1–3 g	All	Acc. (%)	85.77	82.51	**68.86**	76.05	83.26	**77.54**	**72.13**	73.95	87.56	81.91	68.93	74.39
		Pr.	0.857	0.842	0.792	0.757	0.828	0.818	0.764	0.717	0.878	0.833	0.475	0.730
		R.	0.858	0.825	0.689	0.761	0.833	0.775	0.721	0.739	0.876	0.819	0.689	0.744
		FM.	0.857	0.814	0.608	0.754	0.830	0.719	0.609	0.720	0.877	0.800	0.562	0.718
	IG>0	Acc. (%)	**92.59**	**84.14**	66.02	**76.21**	**91.94**	76.76	71.97	**74.62**	**92.67**	**83.83**	69.12	74.73
		Pr.	0.928	0.872	0.780	0.767	0.919	0.824	0.798	0.729	0.926	0.867	0.735	0.742
		R.	0.926	0.841	0.660	0.762	0.919	0.768	0.720	0.746	0.927	0.838	0.691	0.747
		FM.	0.925	0.829	0.554	0.747	0.917	0.702	0.604	0.694	0.926	0.819	0.568	0.710
	Top 70%	Acc. (%)	88.34	83.88	66.44	75.37	87.94	77.17	71.92	73.90	90.81	82.74	**68.93**	**74.93**

(continued)

Table 5.4 (continued)

N-gram	Attribute Selection	Measures	Dataset											
			OMD dataset				HCR dataset				STS-Gold dataset			
			Classifier				Classifiers				Classifiers			
			NB	SVM	KNN	C4.5	NB	SVM	KNN	C4.5	NB	SVM	KNN	C4.5
		Pr.	0.883	0.860	0.771	0.754	0.879	0.826	0.797	0.717	0.907	0.849	0.475	0.741
		R.	0.883	0.839	0.664	0.754	0.875	0.782	0.719	0.738	0.908	0.827	0.689	0.749
		FM.	0.882	0.828	0.564	0.740	0.871	0.732	0.601	0.719	0.908	0.808	0.562	0.717

Table 5.5 Relative improvement in accuracy of classifiers depending on attribute selection options

Dataset		OMD dataset		HCR dataset		STS-Gold dataset	
Attribute selection	IG >0	Top 70%	IG >0	Top 70%	IG >0	Top 70%	
N-gram	Classifiers						
Unigram	NB	**+7.15**	**+8.62**	**+7.75**	**+10.66**	**+6.25**	**+9.30**
	SVM	+1.68	+1.32	−4.11	−0.62	+0.26	+0.05
	KNN	+0.89	+0.26	+0.83	0	+4.47	0
	C4.5	+0.63	−0.47	+0.94	−0.36	0	−0.25
Bigram	NB	**+13.76**	**+7.78**	**+6.97**	**+5.36**	**+16.47**	**+10.23**
	SVM	−2.94	+1.26	+0.36	+0.16	−4.42	−0.73
	KNN	+0.63	0	0	+0.01	+0.05	0
	C4.5	−0.94	−0.21	+0.26	+0.41	−0.29	+0.05
1–3 g	NB	**+6.82**	+2.57	**+8.68**	+4.68	**+5.11**	+3.25
	SVM	+1.63	+1.37	−0.78	−0.37	+1.92	+0.83
	KNN	−2.84	−2.42	−0.16	−0.21	+0.19	0
	C4.5	+0.16	−0.68	+0.67	−0.05	+0.34	+0.54

Figures 5.1, 5.2 and 5.3 illustrate the accuracy level achieved by each classifier based on the pre-processing technique applied to each dataset.

We use three different datasets to evaluate the influence of the dataset domain on preprocessing performance, one with no specific domain tweets and the other with a specific topic. The results of the experiment show that, regardless of the datasets, the effect of preprocessing techniques is the same.

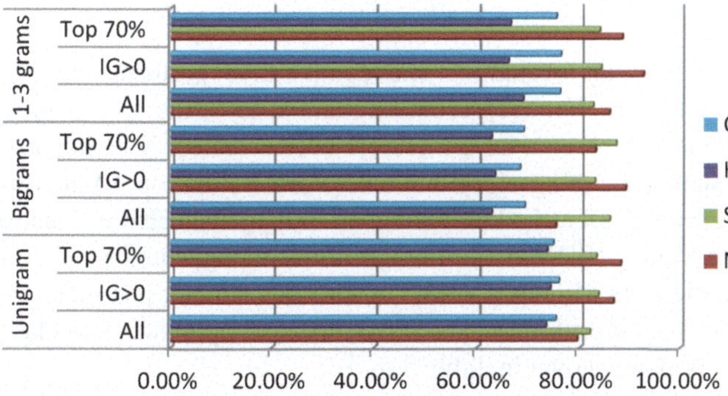

Fig. 5.1 Classifiers' accuracy related with preprocessing techniques on OMD dataset

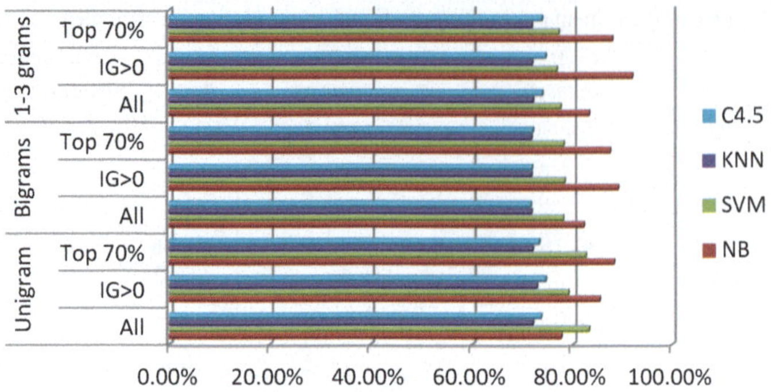

Fig. 5.2 Classifiers' accuracy related with preprocessing techniques on HCR dataset

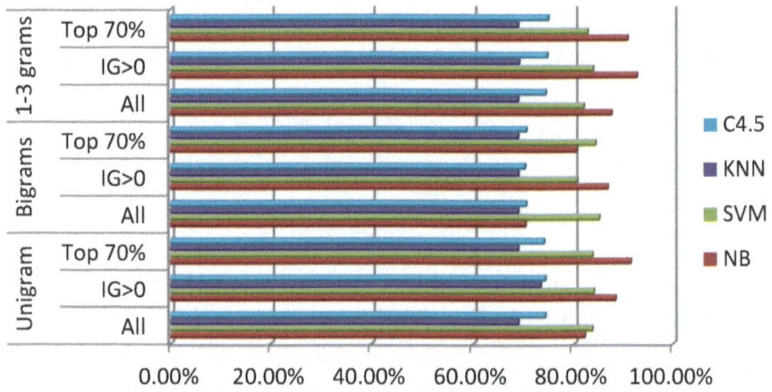

Fig. 5.3 Classifiers' accuracy related with preprocessing techniques on STS-Gold dataset

5.5.4 Evaluation of Stand-Alone Classifiers

This section focuses on evaluating the comparative performance of different approaches to sentiment. Four representative and state-of-the-art machine learning algorithms, provided by Weka, have therefore been selected. Note in particular that the selected machine learning algorithms were among the top 10 most influential data mining algorithms identified in December 2006 by the IEEE International Conference on Data Mining (ICDM), the 11 algorithms implemented by 11 Ants and the component Oracle Data Mining (ODM). In addition, another parameter considered for the choice of the algorithms was to cover different approaches to classification. The classifiers used are shown in Table 5.6.

Naïve Bayes (NB). Naive Bayes classifier is a probabilistic classifier based on applying Bayes' theorem with strong (naive) independence assumptions between the features.

Table 5.6 Experiment classifiers

Classifier	Approach
Naïve Bayes (NB)	Probabilistic learning algorithm
Support Vector Machines (SVM)	Supervised learning model
k-Nearest Neighbor (KNN)	Instance-based learning algorithm
Logistic Regression (LR)	Regression model
C4.5	Decision tree

Support Vector Machines (SVM). A Support Vector Machine (SVM) performs classification by finding the hyperplane that maximizes the margin between the two classes. The vectors (cases) that define the hyperplane are the support vectors.

k- Nearest Neighbors (KNN). The k-Nearest Neighbors algorithm is a instance-based learning, where a case is classified by a majority vote of its neighbors, with the case being assigned to the class most common amongst its k nearest neighbors measured by a distance function.

Decision tree (C4.5). C4.5 is an extension of earlier ID3 algorithm. C4.5 builds decision trees from a set of training data in the same way as ID3, using the concept of information entropy.

A preliminary phase of pre-processing of text and extraction of features is essential for performing the classification. Thus, the tweets were transformed into a vector form for each dataset by applying word tokenization, stemming, and removal of stop-words except for emoticons. In addition, the most valuable and relevant classification attributes were chosen by measuring the class information gain and rejecting it with less than zero information gain. With this option, emerged from the previous section of the evaluation of preprocessing, we can achieve better results of accuracy and reduce training time. The algorithms were then run.

Table 5.7 shows the classification results of the four algorithms used for machine learning. The results show close competition between NB and SVM, as they are more efficient than others, with accuracy rates above 0.8 in all experiments with respective F-measure values, regardless of the dataset. This attests to the fact that in sentiment analysis, NB and SVM classifiers are widespread and the reason they are used in such cases in abundance. On the other hand, the results of the KNN and C4.5 are not so satisfactory, with their accuracy close to 74%.

With respect to the recall, the proportion of positives correctly identified as such, the algorithms return values higher than 0.7 and close to the precision rates in most of our experiments. This also results in a satisfactory F-measure, the weighted harmonic mean of accuracy and recall. In particular, in all dataset classifications, NB and SVM have about 0.8 recall as accuracy. This means the algorithms are highly likely to avoid the classification of false negatives.

Table 5.7 Classification results of machine learning algorithms

Classifiers	Measures	Datasets		
		OMD	HCR	STS-Gold
NB	Accuracy (%)	87.57	85.10	88.69
	Precision	0.876	0.848	0.886
	Recall	0.876	0.851	0.887
	F-Measure	0.873	0.841	0.884
SVM	Accuracy (%)	82.49	79.20	84.10
	Precision	0.836	0.820	0.849
	Recall	0.825	0.792	0.841
	F-Measure	0.814	0.743	0.826
KNN	Accuracy (%)	74.26	73.66	73.77
	Precision	0.773	0.740	0.766
	Recall	0.743	0.737	0.738
	F-Measure	0.704	0.631	0.660
C4.5	Accuracy (%)	74.96	74.35	73.44
	Precision	0.747	0.710	0.715
	Recall	0.750	0.744	0.734
	F-Measure	0.733	0.676	0.680

Each algorithm has been applied in three different datasets in this study. There is no specific domain in STS-Gold while specific topics are addressed by the other datasets. The results show that the comparatively well-performed classifier is the same despite the fact that the performance of the algorithms varies from one dataset to the other. We therefore conclude that, irrespective of the dataset, NB and SVM algorithms are a reliable solution for feeling analysis issues (Fig. 5.4).

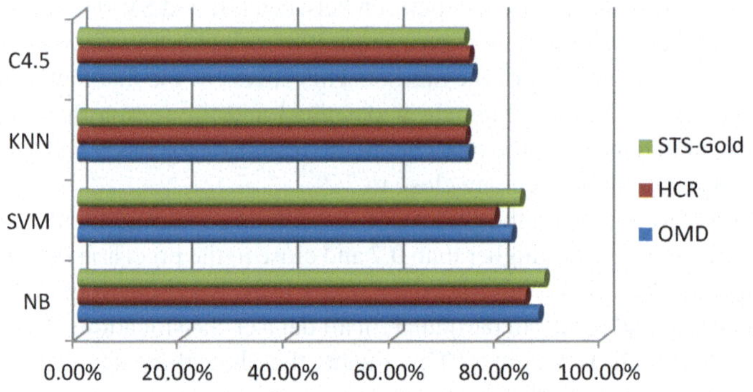

Fig. 5.4 Classification results of machine learning algorithms based on accuracy

The aforementioned results are confirmed also by the research in [19], where different preprocessing techniques were applied to the same datasets.

5.5.5 *Evaluation of Ensemble Classifiers*

Ensemble learning is the process of combining multiple classifiers to achieve greater accuracy than the individual classifiers. We examine four well-known learning techniques for the ensemble in this work. Following, a brief description of each of them is given.

Bagging (Bootstrap Aggregation) generates random training dataset samples and applies basic learning algorithms to each sample. These multiple classifier results are then combined by means of average or majority vote. This method reduces variance in unstable procedures (e.g. decision trees) and thus improves accuracy.

Boosting is an ensemble method which generates complementary models by training each new model on misclassified previous models. The procedure is repeated until the number of models or accuracy of a limit is reached. Although in some cases boosting outperforms bagging, over-fitting of training data is more likely. AdaBoost is this family's most representative algorithm. As weak classifiers, classifiers based on rules, one-two-level trees, neural networks with no hidden layers, etc. are usually used.

Stacking executes various learning algorithms on training data and then uses a meta-classifier that takes each classifier's predictions as additional inputs. This can result in either a bias or variance error decrease depending on the learner used in the combination. Stacking typically delivers better performance than any of the trained models. A logistic regression model with a single layer is often used as a combiner.

Voting is a straightforward adjustment of voting for classifiers for distribution. Various combinations of classification probability estimates are available, such as average probabilities, majority vote, etc. Depending on the combination rule, the ensemble chooses the class that receives the largest total vote.

We try to increase the efficiency of four well-known classifiers in our experiments: Naïve Bayes, SVM, KNN and C4.5, using the ensembles above. In particular, we examine Bagging and AdaBoostM1 algorithms each time using these classifiers as basic learners, Stacking algorithms with these four classifiers and Logistic Regression as the meta-classifier, and Vote algorithms again using the average probability and majority voting combination rule each time using these four classifiers. The number of 10 interactions for the construction of the ensemble models is performed in all implementations, while the pre-processing options mentioned in the previous section have also been applied. Table 5.8 shows the approaches to the ensemble used.

Table 5.8 Ensemble classifiers used

Ensemble type	Implementation
Bagging	Bagging with base classifier: 1. NB 2. SVM 3. KNN 4. C4.5
Boosting	AdaBoostM1 with base classifier: 1. NB 2. SVM 3. KNN 4. C4.5
Stacking	Stacking with 4 base classifiers: – NB – SVM – KNN – C4.5 And meta-classifier: 1. LR
Voting	Vote with 4 base classifiers: – NB – SVM – KNN – C4.5 And combination rule: 1. Average of probabilities 2. Majority voting

Table 5.9 shows the overall picture of all approaches to the ensemble used. In terms of bagging and boosting methods that use a base classifier to optimize their performance, the results show that they return more correctly classified instances than the stand-alone classifier and in most cases boost outperforms bagging. Moreover, when using weak algorithms such as KNN and C4.5, which is confirmed by literature, these approaches yield better accuracy.

Comparing stacking and voting, two methods that combine multiple base classifiers with each individual base classifier, we observe that the weak classifiers, KNN and C4.5 in particular, are outperformed by these ensembles. These ensembles may not necessarily improve the performance of the combination's best classifier; but they certainly reduce the overall risk of misclassification, as it is unlikely that all classifiers will make the same mistake. Table 5.10 is the relative improvement in ensembles' accuracy compared to stand-alone classifiers. It should be noted that KNN and C4.5 outperform more than 10 points in some cases by stacking and voting.

Table 5.9 Classification results of ensemble learning methods

Classifiers		Measures	Datasets		
			OMD	HCR	STS-Gold
NB	Simple	Accuracy (%)	87.57	85.10	88.69
		Precision	0.876	0.848	0.886
		Recall	0.876	0.851	0.887
		F-Measure	0.873	0.841	0.884
	Bagging	Accuracy (%)	**87.74**	**85.10**	88.69
		Precision	0.878	0.848	0.886
		Recall	0.877	0.851	0.887
		F-Measure	0.875	0.841	0.884
	AdaBoostM1	Accuracy (%)	86.34	83.54	**88.85**
		Precision	0.863	0.832	0.887
		Recall	0.863	0.835	0.889
		F-Measure	0.863	0.822	0.885
SVM	Simple	Accuracy (%)	82.49	79.20	84.10
		Precision	0.836	0.820	0.849
		Recall	0.825	0.792	0.841
		F-Measure	0.814	0.743	0.826
	Bagging	Accuracy (%)	81.26	78.68	82.30
		Precision	0.836	0.835	0.842
		Recall	0.813	0.787	0.823
		F-Measure	0.796	0.729	0.799
	AdaBoostM1	Accuracy (%)	81.79	**82.67**	**87.70**
		Precision	0.816	0.820	0.877
		Recall	0.818	0.827	0.877
		F-Measure	0.816	0.814	0.872
KNN	Simple	Accuracy	74.26%	73.66%	73.77%
		Precision	0.773	0.740	0.766
		Recall	0.743	0.737	0.738
		F-Measure	0.704	0.631	0.660
	Bagging	Accuracy	74.61%	73.31%	73.11%
		Precision	0.791	0.671	0.754
		Recall	0.746	0.733	0.731
		F-Measure	0.704	0.623	0.647
	AdaBoostM1	Accuracy (%)	**75.13**	**75.04**	**80.00**
		Precision	0.745	0.722	0.800
		Recall	0.751	0.750	0.800
		F-Measure	0.745	0.695	0.778

(continued)

Table 5.9 (continued)

Classifiers		Measures	Datasets		
			OMD	HCR	STS-Gold
C4.5	Simple	Accuracy	74.96%	74.35%	73.44%
		Precision	0.747	0.710	0.715
		Recall	0.750	0.744	0.734
		F-Measure	0.733	0.676	0.680
	Bagging	Accuracy (%)	**76.53**	75.22	73.11
		Precision	0.768	0.726	0.709
		Recall	0.765	0.752	0.731
		F-Measure	0.749	0.698	0.676
	AdaBoostM1	Accuracy (%)	74.96	**75.22**	**75.25**
		Precision	0.743	0.728	0.741
		Recall	0.750	0.752	0.752
		F-Measure	0.741	0.693	0.712
Combined all	Stacking	Accuracy (%)	**87.57**	**85.10**	**89.02**
		Precision	0.875	0.848	0.890
		Recall	0.876	0.851	0.890
		F-Measure	0.874	0.841	0.886
	Vote-AP	Accuracy (%)	82.49	79.39	84.75
		Precision	0.837	0.803	0.855
		Recall	0.825	0.794	0.848
		F-Measure	0.813	0.753	0.834
	Vote-MV	Accuracy (%)	82.84	79.38	84.75
		Precision	0.840	0.803	0.855
		Recall	0.828	0.794	0.848
		F-Measure	0.818	0.753	0.834

We use three different datasets to evaluate the effect of the dataset domain on ensemble learning, one with no specific domain tweets and the other with a specific topic. The results of the experiment show that, regardless of the dataset, the comparatively well-performed approach is the same.

Figure 5.5 illustrates the accuracy of tested ensembles on the selected datasets.

Table 5.10 Relative improvement in accuracy of ensembles using stand-alone classifiers as baselines

Datasets	Ensembles	Base classifiers			
		NB	SVM	KNN	C4.5
OMD	Bagging	+0.17	−1.23	+0.35	+1.57
	AdaBoostM1	−1.23	−0.7	+0.87	0
	Stacking	0	**+5.08**	**+13.31**	**+12.61**
	Vote-AP	−5.08	0	**+8.23**	**+7.53**
	Vote-MV	−4.73	+0.35	**+8.58**	**+7.88**
HCR	Bagging	0	−0.52	−0.35	+0.87
	AdaBoostM1	−1.56	+3.47	+1.38	+0.87
	Stacking	0	**+5.90**	**+11.44**	**+10.75**
	Vote-AP	−5.71	+0.19	**+5.73**	**+5.04**
	Vote-MV	−5.72	+0.18	**+5.72**	**+5.03**
STS-Gold	Bagging	0	−1.80	−0.66	−0.33
	AdaBoostM1	+0.16	+3.60	**+6.23**	+1.81
	Stacking	+0.33	+4.92	**+15.25**	**+15.58**
	Vote-AP	−3.94	+0.65	**+10.98**	**+11.31**
	Vote-MV	−3.94	+0.65	**+10.98**	**+11.31**

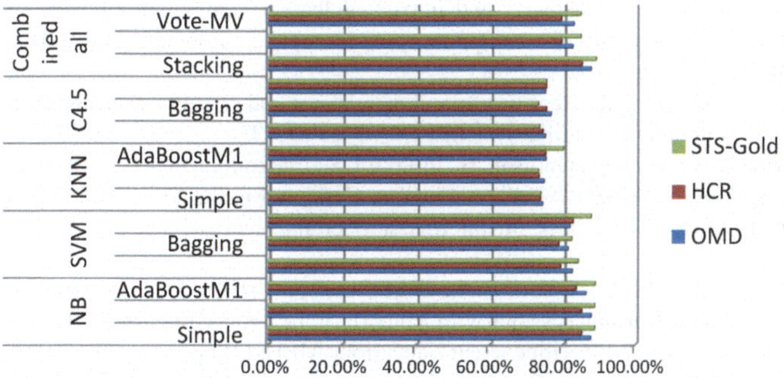

Fig. 5.5 Classification results of ensemble learning methods based on accuracy

References

1. Rodrigo, M., Baker, R., Maria, L., Sheryl, L., Alexis, M., Sheila, P., Jerry, S., Leima, S., Jessica, S., Sinath, T.: Affect and usage choices in simulation problem solving environments. In Proceedings of the 13th International Conference on Artificial Intelligence in Education pp.145–152, Marina Del Ray, CA, USA (2007)
2. Picard, R.W.: Affective Computing. MIT Press, Cambridge (2000)
3. Pinder, C.: Work Motivation in Organizational Behavior. Psychology Press, New York (2008)

4. Calvo, R.A., D'Mello S.K.: Affect detection: an interdisciplinary review of models, methods, and their applications. IEEE Trans. Affect. Comput. **1**, 18–37 (2010)
5. Darwin, C.: The Expression of the Emotions in Man and Animals. Oxford University Press, New York (1998)
6. Ekman, P., Friesen, W.V.: Unmasking the Face: A Guide to Recognizing Emotions from Facial Clues. Malor Books, Los Altos (2003)
7. Ortony, A., Clore, G.L., Collins, A.: The Cognitive Structure of Emotions. Cambridge University Press, New York (1990)
8. Roseman, I.J.: Cognitive determinants of emotion: a structural theory. Rev. Person. Soc. Psychol. **5**, 11–36 (1984)
9. Smith, A., Ellsworth, P.C.: Patterns of cognitive appraisal in emotion. J. Pers. Soc. Psychol. **48**(4), 813–838 (1985)
10. Marsella, S., Gratch, J., Petta, P.: Computational models of emotion. Blueprint Affect. Comput. Sourcebook Manual **11**, 21–46 (2010)
11. Roseman, J., Spindel, M.S., Jose, P.E.: Appraisals of emotion-eliciting events: testing a theory of discrete emotions. J. Pers. Soc. Psychol. **59**(5), 899–915 (1990)
12. D'Mello, S., Picard, R., Graesser, A.: Towards an affect-sensitive Auto Tutor. IEEE Intell. Syst. **22**(4), 53–61 (2007)
13. D'Mello, S.K., Craig, S.D., Fike, K., Graesser, A.C.: Responding to learners' cognitive affective states with supportive and shakeup dialogues. In: Human-Computer Interaction. Ambient, Ubiquitous and Intelligent Interaction, pp. 595–604 (2009)
14. Lehman, B., Matthews, M., D'Mello, S., Person, N.: What are you feeling? Investigating student affective states during expert human tutoring sessions. In: Intelligent Tutoring Systems, pp. 50–59. Springer (2008)
15. Conati, C., Maclaren, H.: Empirically building and evaluating a probabilistic model of user affect. User Model. User-Adap. Interact. **19**(3), 267–303 (2009)
16. Baker, R.S.J.D., D'Mello, S.K., Rodrigo, M.M.T., Graesser, A.C.: Better to be frustrated than bored: the incidence, persistence, and impact of learners' cognitive-affective states during interactions with three different computer-based learning environments. Int. J. Hum. Comput. Stud. **68**(4), 223–241 (2010)
17. Brawner, K., Goldberg, B.: Real-time monitoring of ecg and gsr signals during computer-based training. In: Intelligent Tutoring Systems, pp. 72–77 (2012)
18. D'Mello, S.K., Craig, S.D., Gholson, B., Franklin, S.: Integrating affect sensors in an intelligent tutoring system. In: Affective Interactions: The Computer in the Affective Loop Workshop at 2005 International Conference on Intelligent User Interfaces, 2005, pp. 7–13. AMC Press (2005)
19. Hussain, S., AlZoubi, O., Calvo, R.A., D'Mello, S.K.: Affect detection from multichannel physiology during learning sessions with Auto Tutor. In: Proceedings of the 15th International Conference on Artificial Intelligence in Education, AIED'11, pp. 131–138 (2011)
20. Sabourin, J., Mott, B., Lester, J.C.: Modeling learner affect with theoretically grounded dynamic bayesian networks. In: International Conference on Affective Computing and Intelligent Interaction, pp. 286–295 (2011)
21. Kort, B., Reilly, R., Picard, R.W.: An affective model of interplay between emotions and learning: reengineering educational pedagogy-building a learning companion. In: IEEE International Conference on Advanced Learning Technologies, pp. 0043–0046. IEEE Computer Society (2001)
22. Gee, J.P.: What video games have to teach us about learning and literacy. Comput. Entertainment (CIE) **1**(1), 20–23 (2003)
23. Gee, J.P.: Learning and games. Ecol. Games: Connecting Youth, Games, and Learn. **3**, 21–40 (2008)
24. Kapoor, A., Burleson, W., Picard, R.W.: Automatic prediction of frustration. Int. J. Hum. Comput. Stud. **65**, 724–736 (2007)

25. Rosenzweig, S.: A general outline of frustration. J. Pers. **7**(2), 151–160 (1938)
26. Dollard, J., Miller, N.E., Doob, L.W., Mowrer, O.H., Sears, R.R.: Frustration and Aggression. Yale University Press, New Haven (1939)
27. Morgan, T., King, R.A., Weisz, J.R., Schopler, J.: Introduction to Psychology, 7th edn. McGraw-Hill Book Company (1986)
28. Spector, E.: Organizational frustration: a model and review of the literature. Pers. Psychol. **31**(4), 815–829 (1978)
29. Lazar, J., Jones, A., Hackley, M., Shneiderman, B.: Severity and impact of computer user frustration: a comparison of student and workplace users. Interact. Comput. **18**(2), 187–207 (2006)
30. Graham, S., Weiner, B., Berliner, D., Calfee, R.: Theories and principles of motivation. In: Handbook of Educational Psychology, vol. 4, pp. 63–84 (1996)
31. Hull, C.: Principles of Behavior, 7th edn. Monograph (1943)
32. Heider, F.: The Psychology of Interpersonal Relations. Lawrence Erlbaum Associates, Hillsdale (1958)
33. Weiner, B.: An attributional theory of achievement motivation and emotion. Psychol. Rev. **92**(4), 548–573 (1985)
34. Forsterling, F.: Attributional retraining: a review. Psychol. Bull. **98**(3), 495 (1985)
35. Graham, S.: A review of attribution theory in achievement contexts. Educ. Psychol. Rev. **3**(1), 5–39 (1991)
36. Gagne, M., Deci, E.L.: Self-determination theory and work motivation. J. Organ. Behav. **26**(4), 331–362 (2005)
37. Deci, L., Ryan, R.M.: Self-Determination. Wiley Online Library (2010)
38. Wigfield, A., Eccles, J.S.: Expectancy-value theory of achievement motivation. Contemp. Educ. Psychol. **25**(1), 68–81 (2000)
39. Batool, S., Yousuf, M.I., Parveen, Q.: A study of attribution patterns among high and low attribution groups: an application of Weiners attribution theory. Anthropologist **14**(3), 193–197 (2012)
40. Vockell, E.: Educational Psychology: A Practical Approach. Purdue University, Illinois (2004)
41. Khan, A., Graf, S., Weippl, E.R., Tjoa, A.M.: An approach for identifying affective states through behavioral patterns in web-based learning management systems. In: Proceedings of the 11th International Conference on Information Integration and Web-Based Applications & Services, iiWAS'09, pp. 431–435 (2009)
42. Klein, J., Moon, Y., Picard, R.W.: This computer responds to user frustration: theory, design, and results. Interact. Comput. **14**, 119–140 (2002)
43. Nugent, W.R., Halvorson, H.: Testing the effects of active listening. Res. Soc. Work Pract. **5**(2), 152–175 (1995)
44. Gordon, T.: PET: Parent Effectiveness Training. New American Library, New York (1970)
45. Prendinger, H., Ishizuka, M.: The empathic companion: a character-based interface that addresses users' affective states. Appl. Artif. Intell. **19**(3–4), 267–285 (2005)
46. Hone, K.: Empathic agents to reduce user frustration: the effects of varying agent characteristics. Interact. Comput. **18**(2), 227–245 (2006)
47. Partala, T., Surakka, V.: The effects of affective interventions in human–computer interaction. Interact. Comput. **16**(2), 295–309 (2004)
48. Dweck, S.: Motivational processes affecting learning. Am. Psychol. **41**(10), 1040–1048 (1986)
49. Dweck, S.: Messages that motivate: how praise molds students' beliefs, motivation, and performance (in surprising ways). In: Improving Academic Achievement, pp. 37–60. Academic Press (2002)
50. Wigfield, A., Wentzel, K.R.: Introduction to motivation at school: interventions that work. Educ. Psychol. **42**(4), 191–196 (2007)
51. Shamma, D., Kennedy, L., Churchill, E.: Tweet the debates: understanding community annotation of uncollected sources. In: ACM Multimedia, ACM (2009)

52. Speriosu, M., Sudan, N., Upadhyay, N., Baldridge, J.: Twitter polarity classification with label propagation over lexical links and the follower graph. In: Proceedings of the First Workshop on Unsupervised Methods in NLP, Edinburgh, Scotland (2011)
53. Saif, H., Fernez, M., He, Y., Alani, H.: Evaluation datasets for Twitter sentiment analysis: a survey and a new dataset, the STS-Gold. In: 1st International Workshop on Emotion and Sentiment in Social and Expressive Media: Approaches and Perspectives from AI (ESSEM 2013), Turin, Italy (2013)
54. Bravo-Marquez, F., Mendoza, M., Poblete, B.: Combining strengths, emotions and polarities for boosting twitter sentiment analysis. In: Proceedings of the 2nd International Workshop on Issues of Sentiment Discovery and Opinion Mining, WISDOM 2013 (2014)

Chapter 6
Blending Machine Learning with Krashen's Theory and Felder-Silverman Model for Student Modeling

Abstract This chapter describes briefly the learning and cognitive theories as well as the algorithmic techniques that have been employed for ameliorating social networking-based learning and e-learning. As a testbed for our research, we have designed and fully implemented POLYGLOT, namely a social networking-based intelligent and adaptive tutoring system for the learning foreign languages. POLYGLOT incorporates the following theories, models and techniques for providing a student-centered learning experience. Firstly, it uses the Stephen Krashen's cognitive theory consisting of five hypotheses: the Acquisition-Learning hypothesis, the Monitor hypothesis, the Input hypothesis, the Natural Order hypothesis and the Affective Filter hypothesis. It also employs the Felder-Silverman Learning Style Model, for determining the students' learning styles. POLYGLOT holds a supervised machine learning algorithm (k-nearest neighbors algorithm) which takes as input several students' features, including their age, gender, educational level, computer knowledge level, number of languages spoken and grade on preliminary test, in order to detect automatically their learning style. Moreover, our system incorporates a hybrid machine learning-based approach for error diagnosis, namely Approximate String Matching for conducting natural language processing and diagnosing types of students' errors along with String Meaning Similarity for diagnosing errors due to language transfer interference. POLYGLOT also uses a machine learning-based approach for creating effective groups for students for win-win collaboration. Our system employs a dynamic model for adaptive domain knowledge delivery and personalized assessment units using multiple-criteria decision analysis.

6.1 Employing the Stephen Krashen's Theory of Second Language Acquisition in POLYGLOT

POLYGLOT is an adaptive e-learning system for the tutoring of foreign languages. As such, it is based on the Krashen's theory of second language acquisition [1], consisting of five main hypotheses:

- the Acquisition-Learning hypothesis,
- the Monitor hypothesis,

C. Troussas and M. Virvou, *Advances in Social Networking-based Learning*, Intelligent Systems Reference Library 181, https://doi.org/10.1007/978-3-030-39130-0_6

- the Input hypothesis,
- the Natural Order hypothesis,
- the Affective Filter hypothesis.

The Acquisition-Learning distinction is the most fundamental of all the hypotheses in Krashen's theory. According to the theory, there are two independent systems of second language performance: 'the acquired system' and 'the learned system'. The 'acquired system' or 'acquisition' is the product of a subconscious process very similar to the process children undergo when they acquire their first language. It requires meaningful interaction in the target language—natural communication—in which students are concentrated not in the form of their utterances, but in the communicative act. The "learned system" or "learning" is the product of formal instruction and it comprises a conscious process which results in conscious knowledge 'about' the language, for example knowledge of grammar rules. Towards this direction, POLYGLOT was designed to provide the English and French language concepts and namely the three types of conditionals in both languages in a formal way of instruction, giving the theoretical and grammar rules followed by examples. Moreover, the learning material can be changed by the instructor with the use of POLYGLOT's authoring tool. Furthermore, POLYGLOT supports two different kinds of communication. The first one is the posting on a wall, where all the students can communicate and work on a project with peers or with the instructor. The second way of communication is the asynchronous and instant text messaging between two students or a student and the instructor.

The Monitor hypothesis explains the relationship between acquisition and learning and defines the influence of the latter on the former. The monitoring function is the practical result of the learned concept. According to Krashen, the acquisition process is the utterance initiator, while the learning process performs the role of the "monitor". The "monitor" acts in a planning, editing and correcting function when a specific condition is met, namely the second language learner has sufficient time at his/her disposal in order to think about the correctness of the question provided that s/he has studied the rule. To this direction, POLYGLOT separates the acquisition process from the learning-"monitor" process, by giving to the students the opportunity to learn and be evaluated without time constraints. However, POLYGLOT keeps this information in its log file and uses it when needed. Moreover, POLYGLOT performs the monitoring function by diagnosing students' possible misconceptions and providing assistance when needed. The performance of students depicting the influence of the learning on acquisition is found to be outstanding based on the evaluation results (based on the evaluation results of this research).

The Input hypothesis in Krashen's theory explains how the learner acquires a second language, namely how the second language acquisition takes place. The Input hypothesis is only concerned with 'acquisition', not 'learning'. According to this hypothesis, the learner improves and progresses when s/he receives second language "input" that is one step beyond his/her current stage of linguistic competence. For example, if a learner is at a level "i", then acquisition takes place when s/he is exposed to level "i + 1". To this direction, POLYGLOT holds information about the knowledge level of the student, even from his/her first interaction with the system and performs adaptive actions in order to ensure personalization in the learning process.

The Natural Order hypothesis suggested that the acquisition of grammatical structures follows a "natural order" which is predictable. For a given language, some grammatical structures should be acquired in a proper sequence. This order seemed to be independent of the learners' age, background and conditions of exposure. Indeed, POLYGLOT has a logical gradation of the learning concepts, proceeding from the First type of conditional to the Second etc. As such, this serial presentation of the learning material presents inputs to enhance the progress of the students.

Finally, the fifth hypothesis, that is the Affective Filter hypothesis, embodies Krashen's view that a number of "affective variables" plays a facilitative role in second language acquisition. One important variable for this is the motivation. Krashen claims that learners with high motivation are better equipped for success in second language acquisition. Low motivation can "raise" the affective filter and form a "mental block" that prevents comprehensible input from being used for acquisition. In other words, when the filter is "up", it impedes language acquisition. On the other hand, positive affect is necessary for acquisition to take place. Following this rationale, POLYGLOT incorporates the delivery of motivational messages which can assist the students during their interaction with the system. Moreover, it also detects frustration that can lead to "mental block" and provides motivational messages based on the Attribution Theory, which will be described later in this chapter. Figure 6.1 illustrates the incorporation of the Krashen's theory in POLYGLOT.

More specifically, the Learning Content module in conjunction with the Student Model and the Win-Win Collaboration module are used to support the Acquisition-Learning hypothesis and provide a formal way of instruction and two distinct ways of collaboration, namely posting on wall, instant and asynchronous text messaging and recommendation for effective collaboration. Following, the POLYGLOT Student model and Error diagnosis module are used in order to support the Monitor hypothesis, by providing no time constraints, keeping records in the log file and diagnosing students' misconceptions. The Student model and the Adaptive module are used to support the Input hypothesis by keeping the student's knowledge and performing adaptive actions to students' needs and preferences. Moreover, the Student model and the Learning Content module are used to support the Natural Order Hypothesis by offering logical gradation of the learning concepts. Finally, the Student model and the Frustration Recognition and Respond module are used to support the Affective Filter hypothesis by delivering motivational messages as a response to student's frustration. The aforementioned hypotheses are primarily presented at Chapter 3 and will be further explained afterwards.

6.2 POLYGLOT Learning Content

The domain knowledge of POLYGLOT consists of the grammar phenomenon both in the English and in French language.

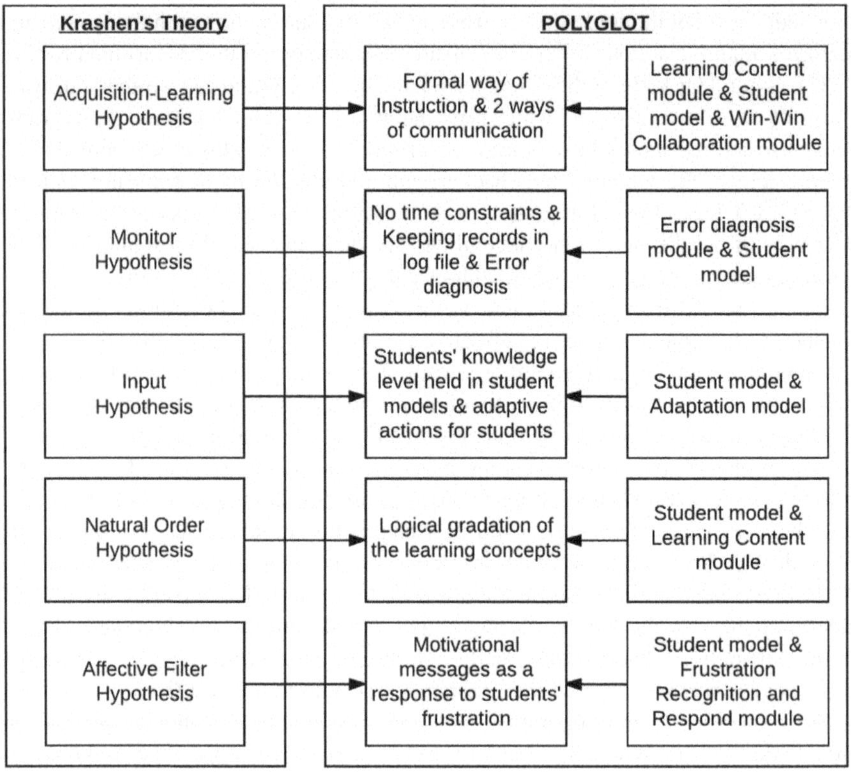

Fig. 6.1 Incorporation of Krashen's theory in POLYGLOT

The full conditional sentences in both languages consist of condition clauses specifying a condition or hypothesis, and a consequence clause or apodosis specifying what follows from that condition. The condition clause is a dependent clause, most commonly headed by the conjunction if, while the consequence is contained in the main clause of the sentence.

Different types of conditional sentences (depending largely on whether they refer to a past, present or future time frame) require the use of particular verb forms (tenses and moods) to express the condition and the consequence. In both languages teaching, the most common patterns are referred to as first conditional, second conditional and third conditional.

Each student can be taught each type of conditionals of both foreign languages in a logical row, which can be depicted in a hierarchical tree. However, as will be shown below, the way of learning is tailored to each student's preferences based on his/her learning style according to the Felder and Silverman model. Particularly, if the student is sequential, then s/he will be given each chapter after the successful completion of the former. If the student is global, then all the chapters are delivered to him/her at his/her first interaction with the system. In both cases, the serial learning can be followed. Also, the hierarchy of this tree depicts the sequencing of levels of the domain concepts of the learning material. The creation of the hierarchy is based on the aforementioned Krashen's model. For instance, the teaching of the first type of conditional precedes the teaching of the third type of conditional which presupposes the learning of the second type of conditionals.

The hierarchy of the domain concepts of the POLYGLOT's learning material is depicted in Fig. 6.2.

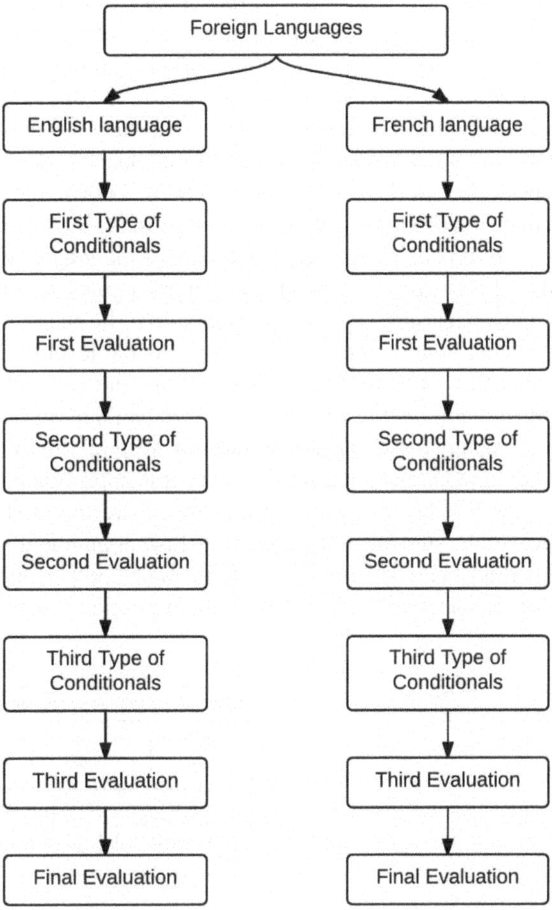

Fig. 6.2 Hierarchy of POLYGLOT domain concepts

6.3 POLYGLOT Student Model

POLYGLOT holds a student model which is responsible for adapting the learning content to the student using machine learning techniques, diagnosing the nature of student's misconceptions along with providing advice to them, when necessary.

More specifically, the student model of POLYGLOT holds static information about the students, and namely their age, gender, level of education, computer knowledge level, proneness to language learning or the foreign languages that they already know, the knowledge level of the student and the learning style in which each student belongs (Fig. 6.3). Furthermore, it holds dynamic information such as their errors and misconceptions along with their progress, being deduced by the interaction between the student and the system. To this direction, POLYGLOT utilizes a multitier student model derived from the theory of the overlay models in conjunction with a multi-dimensional model (8 dimensions) derived from the theory of stereotypes. The overlay model represents the knowledge of the student, while the first dimension of the stereotype model represents the knowledge level of the student, the second dimension represents the type of the language learning misconceptions, the third dimension is the previous foreign language knowledge or proneness to foreign language learning, the fourth dimension is the age of the student, the fifth dimension is his/her gender, the sixth dimension is the level of education, the seventh dimension is the computer knowledge level and the eighth dimension is the learning style. Given that the representation of the student's mastery on a specific learning content is a crucial characteristic in a tutoring system, the overlay technique was chosen to model the learner's knowledge since it is appropriate for that.

The first layer of the aforementioned overlay model is related to the knowledge level of the student, as it results from his/her interaction with the system. The value of this model can be "novice", "intermediate" or "advanced", according to the ACTFL (American Council on the Teaching of Foreign Languages) Proficiency Guidelines[1]

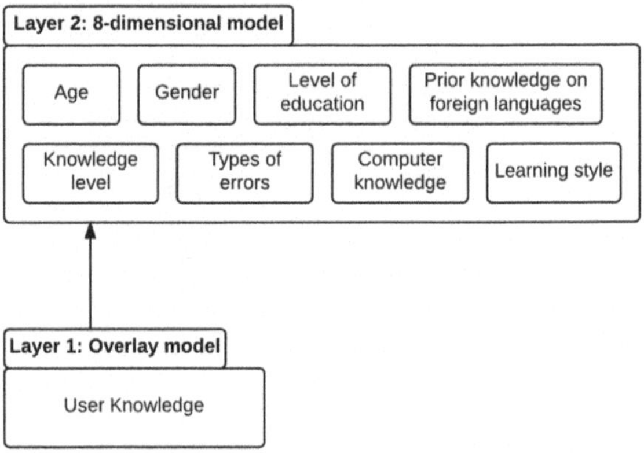

Fig. 6.3 POLYGLOT's student model

[1]https://www.actfl.org/publications/guidelines-and-manuals/actfl-proficiency-guidelines-2012.

and Leung and Li [2]. Novice users are the ones who lack fundamental knowledge of the curriculum being taught. Intermediate users are the ones who have basic understanding of the curriculum while the advanced users can be seen as masters of the curriculum. The guidelines arc broken up into different proficiency levels, such as novice, intermediate and advanced. ACTFL provides a means of assessing the proficiency of a foreign language speaker.

However, defining the learner's knowledge level is not adequate in order to model individual students' needs and abilities. Towards this direction, POLYGLOT can perform misconception detection and diagnosis so that the student model can hold such kind of information. The types of misconceptions are of the following categories [3–5]:

- Accidental slips
- Pronoun mistakes
- Spelling mistakes
- Verb tense mistakes
- Language transfer interference

More specifically, accidental slips are occasional actions, which are not systematic and which the learner himself/herself can correct. For example, the student may have deleted some words and may have forgotten to complete the sentence. In the Table 6.1, a sample of accidental slips is shown.

The pronoun mistakes concern the improper handling of the person. The person refers to the differences among the person speaking (first person), the person spoken to (second person), and the person or thing being spoken about. The common pronoun errors are related to the inappropriate shift in person or in number and the use of the wrong form of a pronoun or the wrong pronoun, being confused when the pronoun is part of a compound subject or object. Table 6.2 gives an insight to this error category.

A spelling mistake occurs when the user has typed the expected word so that one letter is redundant or missing or two neighboring letters have been interchanged. For example, the student has typed "fahter" instead of "father". Table 6.3 provides examples concerning the spelling mistakes.

Table 6.1 Sample of accidental slips

Accidental slips			
Answer with accidental slip	She would have had two laptops if she had digned the contract. (sign)	You would save energy if you sqitched off the lights more often. (switch)	If we had read the book, we would have unferstood the film. (understand)
Answer without accidental slip	She would have had two laptops if she had signed the contract. (sign)	You would save energy if you switched off the lights more often. (switch)	If we had read the book, we would have understood the film. (understand)

Table 6.2 Sample of pronoun mistakes

Pronoun mistakes			
Answer with pronoun mistakes	If she had worn a lighter jacket, the car driver would have seen you earlier. (wear)	You would have watched TV tonight if Peter hadn't bought the theatre tickets for us. (watch)	Them might have arrived on time if they hadn't missed the train. (might arrive)
Answer without pronoun mistakes	If you had worn a lighter jacket, the car driver would have seen you earlier. (wear)	We would have watched TV tonight if Peter hadn't bought the theatre tickets for us. (watch)	They might have arrived on time if they hadn't missed the train. (might arrive)

Table 6.3 Sample of spelling mistakes

Spelling mistakes			
Answer with spelling mistakes	If it rianed, we wouldn't be in the garden. (rain)	I could scor better on the test if the teacher explained me the grammar once more. (can score)	If he greew his own vegetables, he wouldn't have to buy them. (grow)
Answer without spelling mistakes	If it rained, we wouldn't be in the garden. (rain)	I could score better on the test if the teacher explained me the grammar once more. (can score)	If he grew his own vegetables, he wouldn't have to buy them. (grow)

The verb tense mistakes occur when using tenses in a wrong way. For example, the user may have typed "been" instead of "being". Table 6.4 shows examples of this category.

This type of errors when the student uses his/her knowledge and experience in a foreign language as a means of organizing the second language. In POLYGLOT, students can learn two foreign languages, namely English and French; as such, there

Table 6.4 Sample of verb tense mistakes

Verb tense mistakes			
Answer with verb tense mistakes	If it rained, the boys will not play hockey. (rain)	Wouldn't you go out more often if you have to see some friends? (have to see)	She would have yawned the whole day if she has stayed up late last night. (stay)
Answer without verb tense mistakes	If it rains, the boys will not play hockey. (rain)	Wouldn't you go out more often if you had to see some friends? (have to see)	She would have yawned the whole day if she had stayed up late last night. (stay)

Table 6.5 Sample of language transfer interference mistakes

Language transfer interference			
Answer with Language transfer interference mistakes	If you sait a minute, I'll come with you. (wait)	If we arrived at 10, we would mise Tyler's presentation. (miss)	If I went anywhere, it would beau New Zealand. (to be)
Answer without Language transfer interference mistakes	If you wait a minute, I'll come with you. (wait)	If we arrived at 10, we would miss Tyler's presentation. (miss)	If I went anywhere, it would be New Zealand. (to be)

is the possibility of being confused when learning these two languages at the same time. Table 6.5 provides example, concerning the transfer between the two languages that can lead to mistakes.

6.3.1 Approximate String Matching for Error Diagnosis

In order to successfully recognize one or more of the aforementioned categories of errors, POLYGLOT incorporates two algorithmic approaches, as illustrated in Fig. 6.4. The first technique is the Approximate String Matching and tries to find string similarities by matching a student's given "exact" wrong answer with the systems correct stored answer. This technique is responsible for finding strings that match a pattern approximately. The problem of approximate string matching is typically divided into two sub-problems: finding approximate substring matches inside a given string and finding dictionary strings that match the pattern approximately. If string

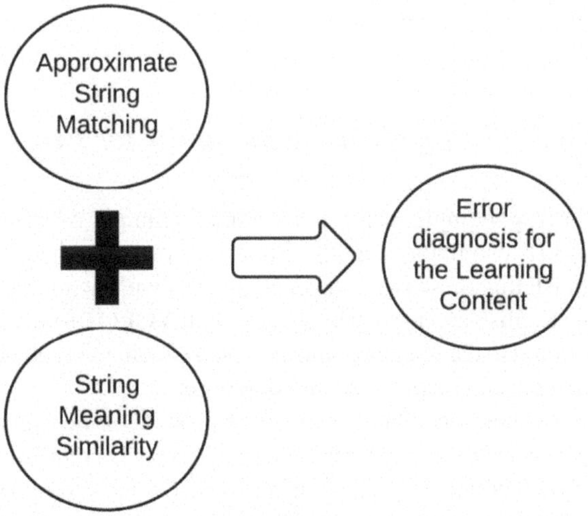

Fig. 6.4 Hybrid model for error diagnosis

matching occurs in a high percentage, POLYGLOT decides whether the mistake lies among the categories of accidental slips, pronoun mistakes, spelling mistakes or verb mistakes.

The closeness of a match is measured in terms of the number of primitive operations necessary to convert the string into an exact match. This number is called the edit distance between the string and the pattern. The usual primitive operations are:

- insertion: For example, the student may have typed cooat, instead of the coat.
- deletion: For example, the student may have typed cot, instead of coat.
- substitution: For example, the student may have typed cost, instead of coat.

POLYGLOT employs the following formulation of the problem: for each position j in the text $T = t_1 t_{2...} t_n$ and each position i in the pattern $P = P_1 P_{2...} P_m$, it computes the minimum edit distance between the i first characters of the pattern, P_i, and any substring $T_{j,j}$ of T that ends at position j.

For each position j in the text T, and each position i in the pattern P, POLYGLOT goes through all substrings of T ending at position j, and determines which one of them has the minimal edit distance E(i, j) to the i first characters of the pattern P. After computing E(i, j) for all i and j, it finds the solution, which is the substring for which E(m, j) is minimal (m being the length of the pattern P).

Computing E(m, j) coincides with the computing of the edit distance between two strings. In fact, POLYGLOT uses the Levenshtein $\gamma\iota\alpha$ E(m, j); the only difference is the initialization of the first row with zeros, and saving the path of computation, that is, whether we used $E(i-1, j)$, $E(i, j-1)$ or $E(i-1, j-1)$ in computing E(i, j).

In the array containing the E(x, y) values, POLYGLOT then chooses the minimal value in the last row, let it be $E(x_2, y_2)$, and follow the path of computation backwards, back to the row number 0. If the field it arrived at was $E(0, y_1)$, then $T[y_1 + {}_1] \ldots T[y_2]$ is a substring of T with the minimal edit distance to the pattern P.

Furthermore, POLYGLOT knows if a learner has proneness in learning foreign languages in order to be able to distinguish if an error occurs due to non-learning or due to confusing by another language.

6.3.2 String Meaning Similarity for Error Diagnosis

Correspondingly, using the second technique of string meaning similarity, POLYGLOT also finds meaning similarities between the given and the correct answer by translating these two answers to the system's available supported languages, namely the English and French languages. POLYGLOT follows the same rationale, as before, tailored to the meaning similarities. As such, the type of Language Transfer Inference mistake can be detected and diagnosed.

As mentioned before, the learning style of the users are detected using the Felder Silverman Learning Style Model. POLYGLOT can infer about the way with which the student prefers to process information (active and reflective learners) and the student progress towards understanding (sequential and global learners). More specifically, active learners can learn by working with others while reflective learners can learn by working alone. Hence, on the one hand active learners want to be able to collaborate with peers in an instant or asynchronous way using the POLYGLOT platform and on the other hand reflective learners do not want to collaborate. Concerning the sequential learners, they prefer to learn in a linear, orderly way in small incremental steps. This process is called "grain size instruction". In this way, the students are given the theory chapter by chapter; after they have learnt the first chapter and been examined for it, they can proceed to the next chapter and so on. On the contrary, global learners are keen on a holistic approach and learning in large leaps. Hence, POLYGLOT gives them the opportunity to have all the chapters available and learn them in the way they prefer.

6.4 Automatic Detection of Learning Styles Based on Felder-Silverman Model Using the K-NN Algorithm

POLYGLOT uses the Felder Silverman Learning Style Model. As mentioned before, the students can be characterized as Active or Reflective learners and Global or Sequential learners. Active learners like to collaborate with peers while reflective learners prefer working alone. Sequential learners like to be taught in linear steps, and each step should follow logically the previous one. Global learners prefer to have available all the learning material and to study in their own pace. To this direction, POLYGLOT offers the capability of collaboration and recommendation for collaboration to active students, while reflective students are given recommendation for collaboration if they ask for it. Also, sequential learners are given the learning material in a grain-size form, from chapter to chapter and they can proceed to the next chapter only if they have successfully completed the previous one. Finally, global students have the capability to navigate through the POLYGLOT's learning material in their own pace.

POLYGLOT offers two ways for the detection of students' learning style. The first one is the traditional way which is conducted by answering the questions proposed by the Felder Silverman questionnaire to detect the aforementioned dimensions. Apart from the completion of questionnaires, they give personal information and namely their age, gender, level of education, computer knowledge level, proneness to language learning/the foreign languages that they already know and to answer a preliminary test to provide their current knowledge level. The second way is the automatic one. POLYGLOT asks the student, who registers, to provide the aforementioned personal information and to answer the preliminary test. After that, POLYGLOT employs machine learning techniques to detect the learning style of the student in order to adapt the learning environment to him/her. The machine learning algorithm, that is used, is the k-nearest neighbor algorithm (k-NN). K-NN was selected for this research since it is one of the top ten data mining algorithms, according to Wu et al. [6].

The k-nearest neighbor algorithm (k-NN) is a non-parametric method used for classification. The input consists of the k closest training examples in the feature space. In k-NN classification, the output is a class membership. An object is classified by a majority vote of its neighbors, with the object being assigned to the class most common among its k nearest neighbors (k is a positive integer, typically small). For example, if k = 1, then the object is simply assigned to the class of that single nearest neighbor. As mentioned above, POLYGLOT uses the following student characteristics in order to employ k-NN and detect his/her learning style (Fig. 6.5):

- age
- gender
- proneness to foreign language learning/number of known languages
- educational level
- computer skills level
- preliminary test score

The aforementioned students' characteristics are used given that they are important for e-learning reasons according to Nakayama et al. [7] and van Setersa et al. [8]. POLYGLOT assigns a weight to the contributions of the neighbors, so that the nearer neighbors contribute more to the average than the more distant ones. For example, as a weighting scheme, POLYGLOT gave each neighbor a weight of 1/d, where d is the distance to the neighbor. As such, students who are nearer to other students, possibly have the same preferences.

The neighbors are taken from a set of objects for which the class is known. This can be thought of as the training set for the algorithm in POLYGLOT.

POLYGLOT makes a prediction on the learning style of a given student using k-NN. The algorithm first calculates the test subjects (student being predicted) similarity to all instances in the training set and finds the k most similar ones. Similarity is calculated with a simple Euclidean distance between the features of the test subject and corresponding features of each instant in the training set. Specifically, the distance measure is given by the formula:

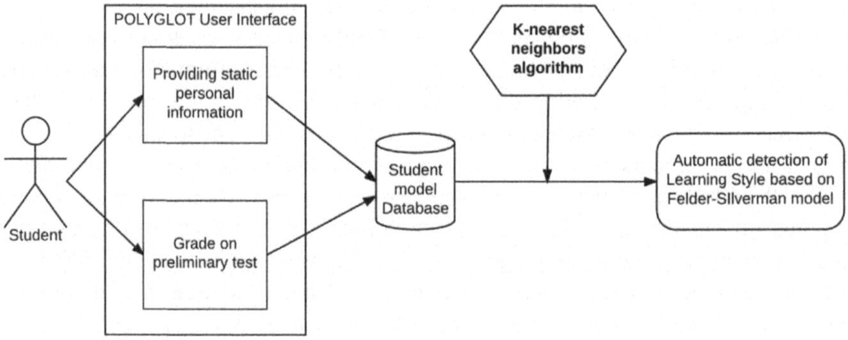

Fig. 6.5 K-nearest neighbors algorithm for automatic learning style detection

$$d(x,y)^n = \sum_{k=1}^{n} (\mathrm{xk} - \mathrm{yk})^\wedge 2$$

where n is the number of dimensions (attributes) and xk and yk are the kth attributes (components) of data objects x and y, respectively.

An example of operation of the automatic detection of a student's learning style using k-NN is the following. Student A has provided to POLYGLOT static information, namely his/her age, gender, number of languages that speaks educational level, computer skills level and the score in the preliminary test. This vector is compared to the students' characteristics of the training set. As such, the student acquires the same learning style with the nearest students.

In the classification phase, k is a priori set to be equal to four. The reason is because of the fact that POLYGLOT wants to detect four distinct learning styles of students and namely:

- Active and Global learners
- Active and Sequential learners
- Reflective and Global learners
- Reflective and Sequential learners

Summarizing, POLYGLOT makes the following steps in order to detect the learning styles of the learners:

- Set K equal to $2^4 = 16$
- Calculate the Euclidean distance
- Determine distance neighbours
- Gather category Y values of nearest neighbours
- Use simple majority of nearest neighbours to predict the value of the query distance.

Finally, it needs to be noted that POLYGLOT has used a training set, namely a vector in a multidimensional feature space, each with a class label; all these vectors where primary users of the e-learning system who served as a training way of k-NN so that it detect the learning style of the students of the private school of foreign languages. The training phase consisted only of storing the feature vectors and class labels of the training samples. The training set consisted of about 100 users, ranged from the age of 11 to 60 years old.

6.5 Tailored Assessments

Due to the growing need to apply intelligence and personalization in digital learning, intelligent adaptive learning systems are rapidly emerging. These data-adaptive solutions are designed to enable separate learning at an individualized tutoring level. Adaptive learning systems are designed to adjust dynamically to the level or type of course content based on the abilities, needs or skills of the personalized student, in ways that accelerate the performance of the learner through both automated procedures and the interventions of the instructors [9]. The purpose of these systems is to use skills and determine what a student really knows and to move students accurately and logically to predefined learning outcomes and skill mastery through a sequential learning path. In view of the above, by providing a student-centered design, adaptive systems have the potential to shift education for students' sake.

Testing is a significant factor in adaptive learning systems. Adaptive tutoring systems in particular use automated processes of student evaluation approaches for self-assessment, diagnosis, and formative evaluation. The more adaptive the tests are to the needs of the learners and the more accurate and effective they are. The digital area and adaptive learning systems facilitate the creation of adaptive testing processes, reducing the time and effort required to create such tests.

In literature review, adaptive tests are called Computer Adaptive Tests (CAT) [10], which are automatically created by a computer system or application. Computer-adaptive testing is designed to adjust their difficulty level to match a student's knowledge and ability based on the responses provided. Considered to be at the forefront of evaluation technology, computer-adaptive testing is an attempt to measure individual students ' abilities more accurately while avoiding some of the issues often associated with the traditional nature of standardized testing. Computer-adaptive testing for students offers a shorter testing session with a smaller number of questions as only those questions that are considered suitable for the student are offered. On the other hand, evaluation creators must create a larger pool of test items so that test systems have sufficient questions to match the varied capabilities of all students taking the exam. Typically, the most current forms of computer-adaptive testing are administered online, and due to computerized scoring, teachers and students can get test results faster than paper-and-pencil testing.

The construction of an algorithm for a particular learner that selects the best exercise/question/activity for a large pool of test objects is a difficult process requiring the identification and auditing of many criteria that concern the abilities and requirements of a student. Such criteria include: level of knowledge, related knowledge or previous experience, preferences for learning and learning method, learning objectives and so on. Thus it seems that multifaceted decision-making analysis (MCDA) [11] is ideal for deciding which test items best suit individual learners ' learning needs and qualifications. In decision making, MCDA explicitly assesses multiple contradictory criteria. There are a range of MCDA approaches and approaches [12]. The weighted sum model (WSM) is used for the presented method of creating an adaptive test. This is because WSM is the most well known and most straightforward MCDA method for evaluating certain alternatives based on various criteria of decision [13]. Multiple benefit criteria are used by WSM and the weight of each alternative is calculated.

Computerized adaptive testing is a key factor in adaptive e-learning systems as it promotes personalized learning; however, it has been handled slightly by the related scientific literature. Many e-learning systems used adaptive computer testing for the practice and/or evaluation of learners [14]. In [15], the authors introduced a new item selection algorithm that integrates knowledge of experts modeled on fuzzy linguistic information that enhances testing adaptation to the student's level of competence. In [11], the authors proposed a framework for assessing, comparing and enhancing the effectiveness of competency indicators in the various publications for primary school teaching materials based on different viewpoints. Their goal was to select the aspired intelligent materials teaching assessment systems. In [12], the authors proposed to use the analytical network process and fuzzy cognitive maps to approach the problem of primary school selection. In [16], the authors described conceptual and technical bases for using computerized adaptive testing to evaluate the skills of modelers to operationalize more rigorous modeling skills assessment. In [10], the authors presented a method for providing the examinee with adaptive tests and useful feedback, which would be used for each French citizen in a platform for certifying digital skills. In [17], the authors proposed a computerized adaptive testing architecture with arbitrarily complex item types, with a particular focus on integration into online learning settings. This architecture is designed in such a way that technology-savvy users can integrate adaptive testing into their online learning platform. In addition, the development and implementation of an adaptive testing system is presented in [18] to support multiple evaluation functions and various devices. In [19], a method for adaptive selection of test questions is presented in a web-based educational system according to the individual needs of students. However, the particular method does not concern the features of specialized learners such as learning style, nor does it combine the features of different learners that imply the ability and needs of the learner in a more representative manner. It also does not use an analysis of decision by multiple criteria. After a thorough investigation in the related scientific literature, however, we came up with the result that there was no implementation of an adaptive assessment framework taking into account the criteria of multiple students along with the types of exercises and the desirable learning objective using multi-criteria decision analysis and the weighted sum model.

6.5.1 Criteria for Tailored Assessment

The presented framework for creating adaptive tests takes into consideration multiple students' criteria along with the types of exercises and the desirable learning objective. The criteria are presented below:

(a) Current Knowledge level of the student in the taught concept: The knowledge level of the student in the taught concept is a very significant characteristic that an adaptive tutoring system takes into consideration in order to deliver the proper learning material to each student. It needs to be underlined that a tutoring system that delivers tailored assessment to students in a dynamic way should respect students' specific needs, such as the advancement of his/her knowledge.
(b) Prior knowledge associated with the taught domain: The prior knowledge of the student can have an effect on his/her current knowledge level as well as the difficulty of the assessment unit. For example, if a student has prior knowledge

in the domain being taught, then the assessment unit which is going to be delivered to him/her should be include the evaluation of more complex units. In the field of language learning that investigates this research, if a student has prior knowledge in tenses, the system will deliver assessment unit related to irregular verbs or passive voice. Also, the English learning process for a learner can be affected by her/his previous knowledge in French or Italian.

(c) Learning style of student: As mentioned above, student presents a different pace and preference of learning being affected by his/her personal needs and interests. Hence, students have a specific style of learning. For example, a student prefers reading a text or looking at images, while another student prefers listening to sounds and voice. For this research and towards tailored assessment, we used several dimensions of the aforementioned Felder-Silverman Learning style model, namely Visual, Verbal and Sensing. These dimensions tend to influence the way of assessment. According to these dimensions, an individual learns better by using one (or two) of the following three ways:

- Visual: a learner, who has a visual learning style, prefers the visual ways of instruction and keeps information better when it is presented in the form of images, charts and/or diagrams. As such, exercises of the tailored test, which will be delivered to the specific student, should involve visualized material, like diagrams, pictures, graphs etc.
- Verbal: a learner, who has a verbal learning style, prefers listening to what is being presented. S/he can take part in discussions and prefers to read audibly the domain to be taught, listening to her/his own voice. Hence, units of the tailored assessment, which will be delivered to the particular student, should provide the possibility to be pronounced, as well as they should give the student the opportunity to record her/his voice while reading the test unit. An example of this is the incorporation of an agent that also reads the exercise.
- Sensing: a learner, who has a sensing learning style, prefers concrete and practical thinking concerned with facts and procedures. S/he prefers experiments and focuses on what can be detected through the five senses. Therefore, the adaptive test can be accompanied with music, as involved by hearing. Moreover, assessment units like crosswords or problem solving with direct visualization of the results seem to be more suitable for this type of students. However, all the exercises of an online test are ideal for such learners as they can utilize keyboard, mouse, or even a tablet to solve them.

It needs to be noted that the learning style of a learner may not always be the same for some tasks. The learner may prefer one learning style for one task, and a combination of others for a different task. An adaptive educational environment or application has to be able to identify each time the preferred learning style for each individual.

(d) Type of assessment units: There are many different types of assessment units that can be included into a test. Types of assessments units can include crosswords, matching exercises, fill in the gaps, multiple choice, right/wrong, etc.

The type of assessment units is closely associated with the difficulty of the exercise, the learning style of the student or to the learning outcome that is aimed to be succeeded. For instance, right/wrong exercises are usually selected for evaluating the student's understanding of the domain knowledge to be taught at the early stages of the learning process, while crosswords and fill in the gaps exercises are selected to be used for assessing more complex issues. The type (or types) of the assessment units that are included into a test are chosen by the instructor.

(e) The intended learning goals according to Revised Bloom taxonomy: The Revised Bloom Taxonomy (RBT) is a taxonomy of educational goals is a framework for classifying the expectations and intentions about the results of the learning process, that is what students will have learnt after their interaction with the tutoring system. It can be used as a way of assisting the building of banks of learning items measuring the same educational goals. Hence, this taxonomy can promote tailored assessment. According to RBT, thinking skills are organized into the following six levels, from the most basic to the more complex levels of thinking:

- Level 1: Remember; namely, recall facts and basic concepts.
- Level 2: Understand; namely, explain ideas or concepts.
- Level 3: Apply; namely, use information in new situations.
- Level 4: Analyze; namely, draw connections among ideas.
- Level 5: Evaluate; namely, justify a stand or decision.
- Level 6: Create: produce new or original work.

As such, according to the learning goal, which implies the knowledge level, the experience and the ability of the learner, the educational system can create tailored assessment units that include a variety of exercises ranging between very simple right/wrong questions (at level 1) or multiple-choice questions asking the learner to select the right answer among a number of given answers occuring after the demonstration of a problem (at level 3) to open-ended questions asking to construct a method or to investigate a way for solving a problem (at level 6). Table 6.6 presents examples of exercises/activities that can be provided to a learner according to the revised Bloom's taxonomy level.

6.5.2 Overview of the Building of the Adaptive Test Algorithm

In this section the algorithm that has been developed for creating adaptive tests using WSM is presented. Log files have to be keeper for both learners and exercises.

- <u>For learners:</u>

- **KnoL**: knowledge level
- **PriK**: prior knowledge on related domain concept

Table 6.6 Examples of exercises according to the revised Bloom's taxonomy

Revised Bloom taxonomy level	Examples of exercises/activities
Remember	Memorizing things; listing things; defining; recognizing facts or procedures
Understand	Identifying, describing and/or classifying; recognizing an option, a fact, a method, a procedure, a situation etc.; explaining facts, things, procedures etc.; select the right answer, method, procedure etc.
Apply	Executing procedures; implementing methodologies and/or theories; solving problems; demonstrating a method
Analyze	Comparing methodologies or solutions; relating information, facts; examining the solutions of a problem; organizing
Evaluate	Giving benefits/mistakes; selecting the most appropriate method or theory to use; udging an action or a problem solution
Create	Planning strategies; designing ways for solving problems; constructing a method; generating new ideas

- **LeaS**: student's learning style
- **RBTL**: Revised Bloom's taxonomy level

The values of the above variables have to be defined dynamically by the adaptive educational system at each interaction of the learner with the system.

- For exercises:

- **EKnoL**: the knowledge level that the exercises concerns
- **APriK**: the extent that the learner's previous knowledge affects the ability of the learner to solve the particular exercise.
- **EVi, EVe, ESe**: The extent that the particular exercise is suitable for learners that have visual, verbal and sensing learning style, correspondingly.
- **AUT**: the type of assessment unit (right/wrong, multiple choice, crosswords, fill in the gap, matching exercise, open-ended questions etc.).
- **ERBTL1, ERBTL 2, ERBTL 3, ERBTL 4, ERBTL 5, ERBTL 6**: The extent that the particular exercise is suitable for the 1st (remember), 2nd (understand), 3rd (apply), 4th (analyze), 5th (evaluate) and 6th (create) Bloom's taxonomy level, correspondingly.

The values of the above variables have to be defined by the instructor for each exercise, when s/he inserts the particular exercise to the bank of exercises. The values of APriK, EVi, EVe, ESe, ERBTL1, ERBTL2, ERBTL3, ERBTL 4, ERBTL5 and ERBTL6 have to be expressed in exactly the same unit, so that the Weighted Sum Model (WSM) can be applied. Therefore, they are expressed as percentage of 100. In other words, they take values between 0 to 100.

- Other notations:

- **S**: the set of exercises that are included into the test
- **p**: the number of exercises of the test
- **exr**: an exercise
- **AUSe**: a set that includes the types of exercises that the instructor desires to be included into the adaptive test
- $\mathbf{w_{PriK}}$: the relative weight of importance of the criterion that is associated with the prior knowledge of the learner in a relative domain concept.
- $\mathbf{W_{LeaS}}$: the relative weight of importance of the criterion that is associated with the learner's learning style.
- $\mathbf{w_{ERBTL}}$: the relative weight of importance of the criterion that is associated with the learner's Revised Bloom's taxonomy level.
- **ELeaS**: The degree that the exercise is suitable for a learner concerning her/his learning style.
- **ERBTL**: The degree that the exercise is suitable for a learner concerning her/his Revised Bloom's taxonomy level.

The steps that the system follows in order to produce the adaptive test:

1. S = Ø
2. The values of p, AUSe, WLeaS and wERBTL are defined by the instructor.
3. The system checks the value of the variable PriK of the learner. If s/he has prior related knowledge, then wPriK = 0, else the system asks the instructor to define the value of wPriK.
4. The system checks the value of the variable PriK of the learner.
5. The system checks the value of the variable LeaS of the learner. If LeaS = 'visual', then ELeaS = EVi; else if LeaS = 'verbal', then ELeaS = EVa; else if LeaS = 'sensing', then ELeaS = ESe.
6. The system checks the value of the variable ERBTL of the learner. If RBTL = 'remember', then ERBTL = ERBTL1; else if RBTL = 'understand', then ERBTL = ERBTL2; else if RBTL = 'apply', then ERBTL = ERBTL3; else if RBTL = 'analyze', then ERBTL = ERBTL4; else if RBTL = 'evaluate', then ERBTL = ERBTL5; else if RBTL = 'create', then ERBTL = ERBTL6.
7. For each exercise exr:

(a) The system checks the value of the variable EKnoL. If EKnoL ≠ KnoL, the system excludes exr from the set S and "goes" to the next exercise, otherwise it proceeds to the next step.

(b) The system checks the value of the variable AUT. If AUT ≠ AUSe, the system excludes exr from the set S and "goes" to the next exercise, otherwise it proceeds to the next step.

(c) Apply the WSM to evaluate the suitability of exr (exrWSM-score) for the particular student:

$$e^{WSM-score} = w_{PriK} * APriK + w_{LeaS} * ELeaS + w_{ERBTL} * ERBTL$$

8. Sort the exercises in descending order based on the WSM scores (if two or more exercises have the same the WSM score, then they are sorted randomly).
9. Insert the first p exercises into the set S (if the selected exercises are less than p, a related message should be displayed to inform the instructor). The test is ready.

References

1. Krashen, S.D.: Principles and Practice in Second Language Acquisition, Monograph, Pergamon Press Inc (1982)
2. Leung, E.W.C., Li, Q.: An experimental study of a personalized learning environment through open-source software tools. IEEE Trans. Educ. **50**(4), 331–337 (2007)
3. Heift T, Schulze, M.: Errors and Intelligence in Computer-Assisted Language Learning: Parsers and Pedagogues, Monograph, Routledge Studies in Computer Assisted Language Learning (2007)
4. Sermsook, K., Liamnimitr, J., Pochakorn, R.: An analysis of errors in written English sentences: a case study of Thai EFL students. Engl. Lang. Teach. (ELT) J. **10**(3), 101–110 (2017)
5. Wu, H.-P., Garza, E.V.: Types and attributes of English writing errors in the EFL context—a study of error analysis. J. Lang. Teach. Res. **5**(6), 1256–1262 (2014)
6. Wu, X., et al.: Top 10 algorithms in data mining. Knowl. Inf. Syst. **14**, 1–37 (2008)
7. Nakayama, M., Yamamoto, Y., Santiago, R.: The impact of learner characteristics on learning performance in hybrid courses among Japanese students. Electron. J. E-Learn. **5**(3), 195–206 (2007)
8. van Setersa, J.R., Ossevoortb, M.A., Trampera, J., Goedhart, M.J.: The influence of student characteristics on the use of adaptive e-learning material. Comput. Educ. **58**(3), 942–952 (2012)
9. Troussas, C., Krouska, A., Virvou, M.: Reinforcement theory combined with a badge system to foster student's performance in e-learning environments. In: Proc. 8th IEEE International Conference on Information, Intelligence, Systems & Applications (IISA), pp 1–6. Larnaca, Cyprus (2017)
10. Vie, J.J., Popineau, F., Tort, F., Marteau, B., Denos, N.: A heuristic method for large-scale cognitive-diagnostic computerized adaptive testing, Proceedings of the Fourth. In: Proc 4th ACM Conference on Learning @ Scale, pp. 323–326. USA (2017)
11. Chen, C.H., Tzeng, G.H.: Creating the aspired intelligent assessment systems for teaching materials. Expert Syst. Appl. **38**(10), 12168–12179 (2011)
12. Baykasoglu, A., Durmusoglu, Z.D.U.: A hybrid MCDM for private primary school assessment using DEMATEL based on ANP and fuzzy cognitive map. Int. J. Comput. Int. Sys. **7**(4), 615–635 (2014)
13. Triantaphyllou, E.: Multi-Criteria Decision Making Methods: A Comparative Study. Kluwer, Norwell, MA (2000)
14. Cisar, S.M., Cisar, P., Pinter, R.: Evaluation of knowledge in object oriented programming course with computer adaptive tests. Comput. Educ. **92**, 142–160 (2016)
15. Badaracco, M., Martinez, L.: A fuzzy linguistic algorithm for adaptive test in intelligent tutoring system based on competences. Expert Syst. Appl. **40**(8), 3073–3086 (2013)
16. Graf, D., Oppl, S., Eckmaier, A.: Towards BPM skill assessment using computerized adaptive testing. In: Proc. 9th Conf. on Subject-oriented Business Process Management, Darmstadt, Germany (2017)

17. Oppl, S., Reisinger, F., Eckmaier, A., Helm, C.: A flexible online platform for computerized adaptive testing. Int. J. Educ. Technol. High. Educ. **14**(2), 1–21 (2017)
18. Huang, Y.M., Lin, Y.T., Cheng, S.C.: An adaptive testing system for supporting versatile educational assessment. Comput. Educ. **52**(1), 53–67 (2009)
19. Barla, M., Bielikova, M., Ezzeddinne, A.B., Kramar, T., Simko, M., Vozar, O.: On the impact of adaptive test question selection for learning efficiency. Comput. Educ. **55**(2), 846–857 (2010)

Chapter 7
Regression-Based Affect Recognition and Handling Using the Attribution Theory

Abstract This chapter presents the automatic frustration detection module and the response to it through motivational messages using a learning theory of our social networking-based language learning system, called POLYGLOT. In POLYGLOT, students can declare their affective state among "happy", "frustrated" and "neutral". However, their interaction with the tutoring system, i.e. experiencing difficulty in a test or receiving a bad grade, can be a blockage of their goal and as such the reason of feeling a negative emotion, such as frustration. Hence, POLYGLOT can detect students' frustration by using the linear regression model. The relationships are modeled using linear predictor functions whose unknown model parameters are estimated from the data. Finally, the POLYGLOT's response on frustration is the delivery of motivational messages based on the Attribution Theory, involving a three-stage process underlying that behavior must be observed/perceived, must be determined to be intentional and is attributed to internal or external causes. With the use of motivational messages, the students are assisted in the educational process and are not willing to quit learning.

7.1 Declaration and Handling of Affective States

Before the student interacts with the system, POLYGLOT asks him/her to declare that he/she is "happy", "frustrated" or "neutral." This feature follows the basic principles of social networking sites (e.g. Facebook) that tend to ask the user how he/she feels. This first step of declaring an impact serves as the threshold for managing students' affective states. Therefore, by delivering messages, the system can primarily support users as being following described. To this end, even before interacting with the system, POLYGLOT can support students and motivate them to reach their goals.

More specifically, before the student starts to study the learning content and be evaluated by POLYGLOT, s/he has the capability to declare his/her affective state. Based on this declaration, POLYGLOT takes this input and provides several messages to the students based on the declared affective state. These messages are not motivational given that there is no need to motivate students since they are not frustrated by the interaction with the ITS. As will be seen in the next Chapter, POLYGLOT

C. Troussas and M. Virvou, *Advances in Social Networking-based Learning*, Intelligent Systems Reference Library 181, https://doi.org/10.1007/978-3-030-39130-0_7

responds to frustration when it takes information by the student model log file, such as students' poor grades, response time on exercises and liking/disliking of questions. To this direction, POLYGLOT employs conditional constructs in order to decide the appropriate message for each declared affective state. Following, Figs. 7.1, 7.2 and 7.3 illustrate samples of messages to students before their integration with POLYGLOT. Specifically, Fig. 7.1 shows a message to students when they are happy, Fig. 7.2

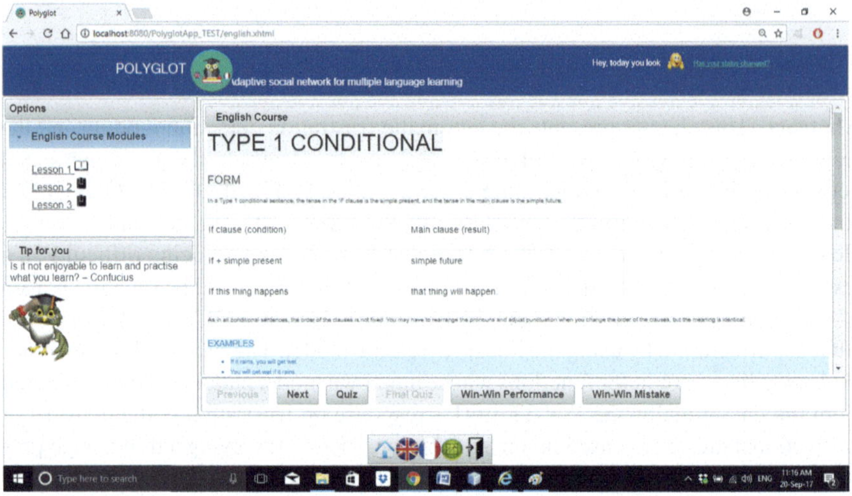

Fig. 7.1 Message to student (before any kind of interaction), when s/he is happy

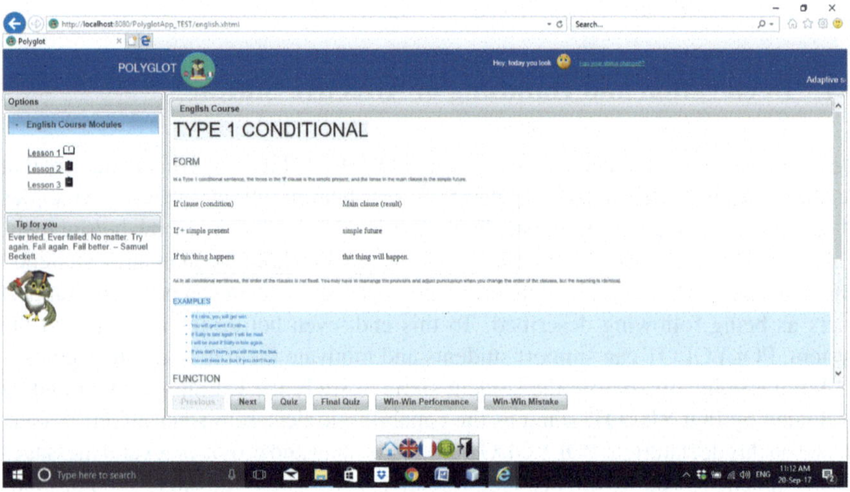

Fig. 7.2 Message to student (before any kind of interaction), when s/he is frustrated

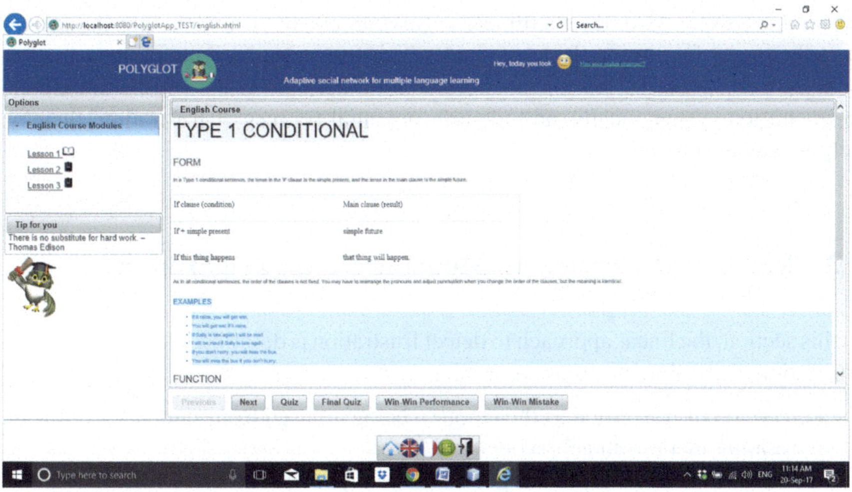

Fig. 7.3 Message to student (before any kind of interaction), when in a neutral affective state

shows a message to unhappy/frustrated students while Fig. 7.3 shows a message to students with a neutral affective state.

7.2 Automatic Detection of Frustration

After the interaction with POLYGLOT, the affective state of the student can change. Hence, POLYGLOT incorporates mechanisms to detect frustration.

Frustration of students is more applicable to computer learning environments [1–7]. In learner-centered affective states, identifying and responding to the negative affective states are significant since it might render the student susceptible to quit learning [8]. According to Gee [9], frustration should be kept below a certain level in order to avoid high stress, powerful anger or intense fear. Moreover, frustration is a cause of student's disengagement and can eventually lead to attrition [10]. For all the above reasons, this dissertation takes into account only students' frustration and responds on this emotion.

POLYGLOT creates a model to detect and respond to frustration accurately and in real-time, when students are working with the ITS. In this chapter, the approach to detect student's frustration, when they interact with the ITS, is described. Then, the way that the approach is applied to POLYGLOT is presented. The frustration model is created by constructing features from the POLYGLOT's log data related to the frustration. The focus is placed on the instances of frustration that occur due to goal blockage. Frustration, in this case, is considered to be a negative emotion, as it interferes with a student's desire to attain his/her goals. To model the frustration, the rectilinear regression classifier is used. The rectilinear regression model is flexible, fast and accurate. Also, the rectilinear regression classifier models into the factors contributing to frustration. It determines which features contribute most to frustration, as well. Thus the rectilinear regression model can be the means to respond to frustration systematically, and identify potential sources of frustration, thereby, helping students to avoid it.

7.3 Rectilinear Regression Model to Detect Frustration

In this section, the linear approach to detect frustration is described. In order to model frustration, the following steps are used:

1. Perception of frustration as the emotion being aroused from students' confusion preventing them from achieving a goal.
2. Identification of the students' goals while they interact with the system (goal1, goal2, ..., goaln).
3. Reporting the blocking factors of each identified goal (block.goal1, block.goal2, ..., block.goaln). Operationalization for POLYGLOT, using its log data.
4. Creation of a rectilinear regression model for frustration index (F_i) with the blocking factors identified.
5. Determination of the weights of the rectilinear regression model using labeled human observation data.

The selection and combination of features from the POLYGLOT's log file is conducted through a systematic process based on an analysis of goal-blocking events. According to Step 1, the goals of the students are identified with respect to their interaction with the ITS, and the top n goals are selected in Step 2. Based on information from the student log files, a blocking factor (block), for each of the n goals is identified (Step 3). For example, the block.goalj represents the blocking factor for the goalj. A linear model for F_i is formulated; F_i represents the frustration index at the *i*th question based on the blocking behaviors of student goals (Step 4). The features in the rectilinear regression model are constructed based on the aforementioned perception of frustration. A threshold is applied to the frustration index Fi in order to detect whether the student is frustrated or not. The average of values used to represent frustration and non-frustration, during the training process, is used as threshold. The weights of the rectilinear regression are determined during the training process (Step 5)—with labeled data from human observation—which is an independent method to identify affective states. The proposed rectilinear regression model to detect frustration is given as follows:

$$F_i = a[w_0 + w_1 * \text{block.goal1} + w_2 * \text{block.goal2} + \cdots + w_n * \text{block.goaln} + w_{n+1} * t_i] + (1 - a)[F_{i-1}] \quad (7.1)$$

The weights w_0, w_1, w_2, ..., w_n in the equation above are determined by the rectilinear regression analysis, which is explained later in this chapter. As explained in the previous paragraph, the terms block.goal1, block.goal2, …, block.goaln, are the blocking factors for goals goal1, goal2, …, goaln, respectively. The term t_i symbolizes the time spent by the student to answer the question i. Lazar et al. [11] state that the time spent to achieve the goal is an important reason of frustration. The last term in the equation, $(1\text{-}a)[F_{i\text{-}1}]$ accounts for the cumulative effect of frustration. We include this term on the basis of Klein et al. [12], which states that frustration is cumulative in nature. The value of a, determines the contribution of frustration at $(i-1)$th question to frustration at ith question; a ranges from 0 to 1. We assume that the student is not frustrated at the beginning of their interaction with the ITS, and hence, choose $F_i = 0$ for $i = 1, 2, 3$. The scope of this approach is to identify frustration that occurs due to students' goal blockage (blocking factors) while interacting with the ITS. Instances of frustration, that might have occurred due to external situations unrelated to the students' interaction with the ITS, are excluded. Hence, the primary concern is the accuracy of the detection (precision), no matter how many the frustration instances are (recall).

7.4 Incorporation of the Rectilinear Regression Model in POLYGLOT

In this section, the application of the rectilinear regression approach to POLYGLOT log data is explained. The goal is to detect frustration of the students while they interact with POLYGLOT. The creation of the rectilinear regression model is based on the following steps:

Step 1. Definition of Frustration:

As mentioned above, the perception of frustration is important for the rectilinear regression model and is related to the emotion being aroused from student's confusion and prevents him/her from achieving a goal. At previous chapters, frustration was defined based on the researches of Lazar et al. [11], Morgan et al. [13] and Spector [14] as follows:

- Frustration is the blocking of a behavior directed towards a goal.
- The distance to the goal is a factor that influences frustration.
- Frustration is cumulative in nature.
- Time spent to achieve the goal is a factor that influences frustration.
- Frustration is considered as a negative emotion, because it interferes with a student's desire to attain a goal.

Step 2. Identification of Students' Goals:

The four most common goals of students, while interacting with POLYGLOT, are identified. According to Daish et al. [15], McWhaw and Abrami [16], Jacob and Rockoff [17], Ewing [18], the students' goal is the achievement of good grades in all the tests of the e-learning system. Based on these researches, we also asked the 80 students from the private school of foreign languages to state which their goals are before using POLYGLOT. Their answers coincide with the aforementioned researches and are the following:

- To get the correct answer to a single question
- To pass successfully the test of each chapter
- To reach the final test (having passed successfully the tests of all chapters)
- To pass successfully the final test.

The corresponding blocking factors of each goal are discussed in the next step.

Step 3. Defining the Blocking Factors:

POLYGLOT involves the goals goal1, goal2, goal3, goal4 and their corresponding blocking factors block.goal1, block.goal2, block.goal3, block.goal4. To model the blocking factor (block) of each goal, several characteristics are taken into account, such as the students' response to questions, the time needed to answer each test and their liking/disliking in questions of chapter tests and final test; these features are being captured in the POLYGLOT's student log file.

Concerning the goal1, namely "to get the correct answer to a single question", the blocking factor is having the wrong answer to this single question. We use a_i to represent the answer of the single question. Specifically, when the answer is correct then $a_i = 1$, and when the answer is wrong then $a_i = 0$. The blocking factor of the goal1 is captured using

$$\text{block.goal1} = (1 - a_i)$$

Concerning the goal2, namely "to pass successfully the test of each chapter", the student should answer correctly all the questions of the test of each chapter. This goal can be blocked, if a student gets a grade which does not allow to successfully pass the test in order to proceed to the next chapter, as a logical sequence of the learning material (even if the student is a global student). Since the blocking factor by getting the wrong answer to the current question is partly addressed in block.goal1, we consider only the blocking factor by achieving more correct answers in order to take requested grade to pass. Hence the block.goal2 has two components. One way in which the goal2 can be blocked is when the student answers correctly some of the needed questions and the majority of them wrongly. Each test of each chapter has 10 questions and as such this is captured by the blocking factor block.goal2 as follows:

$$\text{block.goal2} = \left(a_{i-4} * a_{i-3} * a_{i-2} * a_{i-1} * (1 - a_i)^6\right)$$

Concerning the goal3, namely "to reach the final test" (having passed successfully all the tests of each chapter), the student should answer correctly the majority of the questions in each test of the three chapters for the one foreign language. The same happens correspondingly for the other foreign language. This goal can be blocked, if the student does not achieve the needed grade in any of the three tests. Namely, s/he does not answer correctly the majority of the questions in any of the three tests. This is captures by the block.goal3 as follows:

$$\text{block.goal3} = (\text{block.goal2})_{T_1} + (\text{block.goal2})_{T_2} + (\text{block.goal2})_{T_3}$$

In the above formula, T_1 symbolizes the test 1, T_2 symbolizes the test 2 and T_3 symbolizes the test 3.

Concerning the goal 4, namely "to pass successfully the final test", the student should answer correctly the majority of the questions of the final test. Given that the final test has 30 questions, the blocking factor of goal4 is captured using:

$$\text{block.goal4} = \left(a_{i-12} * \cdots * a_{i-2} * a_{i-1} * (1 - a_i)^{18}\right)$$

Step 4. Employment of the Rectilinear Regression Model:

The mathematical model to detect frustration in POLYGLOT is given in Eq. 7.2, with the individual terms block.goal1, block.goal2, block.goal3 and block.goal4, being defined in the above equations:

$$\begin{aligned} F_i = a[&w_0 + w_1 * \text{block.goal1} + w_2 * \text{block.goal2} + w_3 * \text{block.goal3} \\ &+ w_4 * \text{block.goal4} + w_5 * t_i] + (1 - a)[F_{i-1}] \end{aligned} \tag{7.2}$$

7.5 Respond to Frustration

As mentioned in previous chapters, the followed strategy to respond to frustration consists of the following aspects:

- Create motivational message to attribute the students' failure to achieve the goal to external factors.
- Create messages to praise the students' effort instead of outcome.
- Create messages with empathy, which should make the student feel that s/he is not alone in that affective state.
- Create message to request student's feedback.
- Display messages using an agent.

We create and display the messages to motivate the students based on the reasons for why the student is frustrated. The prime reason for frustration is the goal failure. The possible reasons for goal failure are due to the non-achievement of good grades [15–18] and are identified from the students' goal while they interact with the ITS. We represent these reasons as "events". To create and display the messages we consider the events in POLYGLOT as listed in Table 7.1. The frustration model is modified to identify the Reasons of Frustration (RF) as shown in Eq. 7.3

$$\text{RF} = \text{block.goal1} + \text{block.goal2} + \text{block.goal3} + \text{block.goal4} \tag{7.3}$$

The values of RF and its corresponding reasons for failure are detailed in Table 7.1. The value of RF will be in the range of 0–2. For instance, if the goal1, that is getting

Table 7.1 Events as reasons of goal failures

Event	RF value	Pattern of answers
Ev1	0	Incorrect student's response to a single question a_i
Ev2	1	Incorrect student's response a_i to the majority of questions of the test of a single chapter
Ev3	2	Incorrect student's response a_i to the majority of questions of the tests of all the chapters
Ev4	3	Incorrect student's response a_i to the majority of questions of the final test

the correct answer to a single question, is blocked then it is identified by block.goal1 which is that then answer to a single question is wrong.

7.6 Delivery of Motivational Messages Based on the Attribution Theory

The motivational messages are based on the Attribution Theory and created using the reasons of frustration held in the log data of POLYGLOT. Attribution Theory was proved to be a useful conceptual framework for the study of motivation in educational contexts [19].

As mentioned in earlier chapters, the Attribution Theory is a framework assuming that people try to determine why people do what they do, namely, it interpret the causes to an event or behavior. A three-stage process underlies an attribution:

- behavior must be observed/perceived
- behavior must be determined to be intentional
- behavior attributed to internal or external causes.

The Attribution Theory is mainly about achievement. According to it, the most important factors affecting attributions are ability, effort, task difficulty, and luck. Attributions are classified along three causal dimensions:

- locus of control (two poles: internal versus external)
- stability (do causes change over time or not?)
- controllability (causes one can control such as skills versus causes one cannot control such as luck, others' actions, etc.).

According to the theory, when a student succeeds, s/he attributes his/her successes internally. Namely, s/he believes that success is due to high ability and effort which s/he is confident of. When a rival succeeds, a student tends to credit external (e.g. luck). When a student fails or makes mistakes, external attribution is more likely to be used, attributing causes to situational factors rather than blaming his/her fault. Thus, failure doesn't affect their self-esteem but success builds pride and confidence. When students fail or make mistakes, internal attribution is often used, saying it is due to their internal personality factors. The main principles of the Attribution Theory are the following:

- Attribution is a three-stage process: (1) behavior is observed, (2) behavior is determined to be deliberate, and (3) behavior is attributed to internal or external causes.
- Achievement can be attributed to (1) effort, (2) ability, (3) level of task difficulty, or (4) luck.
- Causal dimensions of behavior are (1) locus of control, (2) stability, and (3) controllability.

In view of the above, motivating the students' success with messages, praising their ability, can enhance students in the learning process and motivating the students' failure with messages which attribute the failure to external or unstable or controllable factors will help them to set a new goal with self-motivation. Figure 7.4 illustrates how the principles of the Attribution Theory are used by POLYGLOT in order to deliver motivational messages. The motivational messages in Fig. 7.4 are a sample of the ones delivered by POLYGLOT and are in the yellow boxes.

The following parameters are identified from the POLYGLOT log data, and are taken into account while creating the motivational messages. The messages are given in Table 7.2 with condition to display the message and the reason for creating the message.

Average Response Time (Resp_T) is the average time taken to answer the questions in POLYGLOT by students. For the test of each chapter, the average response time from POLYGLOT existing log data is 50 s. This time coincides with the time that the teachers of the private school of foreign languages proposed as the average response time. The Response Rate is the percentage of instances when students

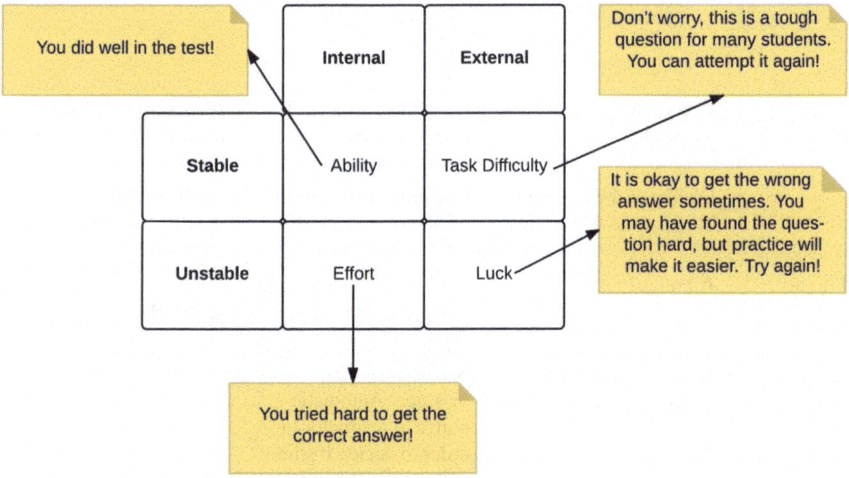

Fig. 7.4 Sample of motivational messages based on the Attribution Theory

Table 7.2 Motivational messages responding to frustration

	Condition/event	Motivational message	Explanation
Fru_ins = 1	Ev1	You did well in the last question!	Stating the reason for frustration and praising the effort of the learner
	Ev2	You did well in the test!	
	Ev3	You did well in all the tests!	
	Ev4	You did well in the final test!	
	Resp_T > Average response time	You tried hard to get the correct answer!	Praising the effort of the learner
	Resp_T < Average response time	Try harder!	Motivating the learner
	Chapter test	For sure, you will do well in the next questions!	
	Final test	You may succeed next time!	
	Dislike of the majority of questions of chapter test	You can take a break and try again with a new state of mind!	
Fru_ins = 2	Final test	Don't worry, this is a tough question for many students. You can attempt it again!	Attributing the failure to the difficulty of the question and motivating the learner
	Response Rate > 50% in the final test	It is okay to get the wrong answer sometimes. You may have found the question hard, but practice will make it easier. Try again!	Sharing the feeling of the learner—showing empathy
	Response Rate < 50% in the final test	It seems that this is a tough question for many students. Try again!	Attributing the failure to the difficulty of the question
	Dislike of the majority of questions of final test	It seems that this test is disliked! Try again!	Attributing the failure to the disliking of the question
Fru_ins = 3	All questions	You can send a message to the instructor if you want!	Receiving student's feedback
	Dislike of questions of chapter test or final test	No motivational message. Automatic notice to instructor in order to decide if s/he will change the question	

answered the question correctly. We calculate the response rate for all questions using the POLYLGLOT's existing log data. We represent the response rate as RR. Frustration instances in a current session are represented by Fru_ins. The Fru_ins counts the number of frustration instances detected in the session, namely how many times the student has faced frustration in a current session.

Wrong answers in the majority of the questions of each chapter test or the final test are kept in the student model log file and used for the delivery of motivational messages. Liking and disliking of exercises serve as a valuable input to POLYGLOT and motivational messages are also delivered to students based on this interaction. The messages, discussed in Table 7.2, are concatenated based on the conditions and displayed to the students. Each message can be appeared in a speech bubble from an agent (an owl, as shown in previous chapters).

For the events listed in Table 7.2, that is for each goal failure, POLYGLOT shows the motivational messages (as shown in Fig. 7.5) based on the student's response time in answering the questions, grade, type of the question (belonging to a chapter test or final test) and liking/disliking of the questions. POLYGLOT's frustration model takes into consideration the following factors:

- For the first instance of frustration, POLYGLOT chooses the message based on the time spent by the student to answer the question, that is, Resp_T. If the student spent more than an average response time then, based on the event, the message praising the student's effort of answering the question will be shown. If the student spent less than an average response time then, the message to motivate the student will

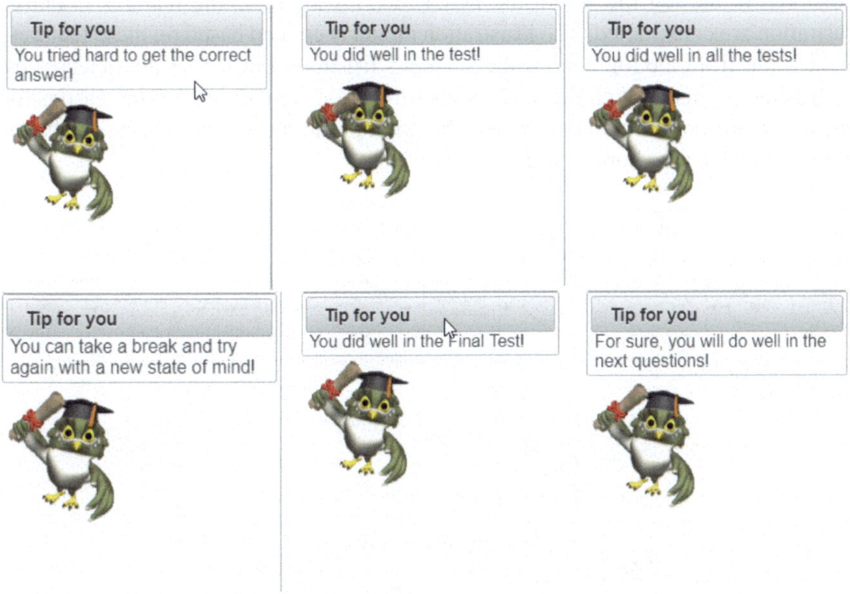

Fig. 7.5 Sample of motivational messages to students (after their interaction with POLYGLOT)

be shown. This is to praise the students' effort to answer the question. Accordingly, POLYGLOT selects the appropriate message in case of a student's disliking of a question of the chapter test.

- For the second instance of frustration, POLYGLOT chooses the message based on the response rate. If the response rate is more than 50% of the average then the message to motivate the student will be shown. If the response rate is less than 50% of the average, then the message to attribute the failure to the difficulty of question will be shown. If the response rate is less than 50% of the average, then this question might be difficult for many of the students. This is to attribute the students' failure to difficulty of the question; hence, the student will be motivated for the next questions. Accordingly, POLYGLOT selects the appropriate message in case of a student's disliking of a question of the final test.
- For the third instance of frustration, the student's feedback is gathered either implicitly or explicitly.

Figures 7.6 and 7.7 illustrate examples of delivering motivational messages to students based on their interaction with POLYGLOT. For instance, student A, who is in the final test, needs a lot of time to answer an exercise, has poor results and has liked the majority of the questions, will receive the message "You may succeed next time!" in case s/he has faced a high number of frustration instances (motivation to the student) or the message "It is okay to get the wrong answer sometimes. You may have found the question hard, but practice will make it easier. Try again!" in case s/he has faced a low number of frustration instances (empathy to the student). As shown in Fig. 7.8, the student's interactions with the user interface of POLYGLOT are stored in the log file. From the POLYGLOT's log data, the features to detect frustration are constructed. The Frustration model is created based on these features, as the input from the log data. If the student's frustration instances are detected by the frustration model, then the reasons for frustration are identified. The reasons for frustration are represented as events. The appropriate motivational message based on the events and the data from log file is selected.

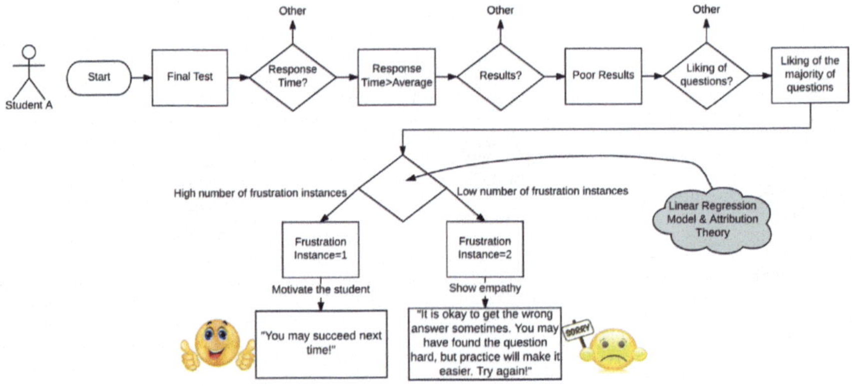

Fig. 7.6 First example of delivering motivational messages to students

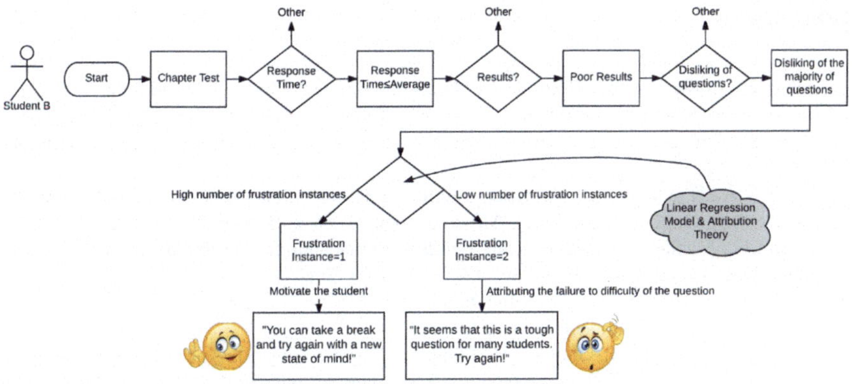

Fig. 7.7 Second example of delivering motivational messages to students

Fig. 7.8 Methodology to detect and respond to frustration in POLYGLOT

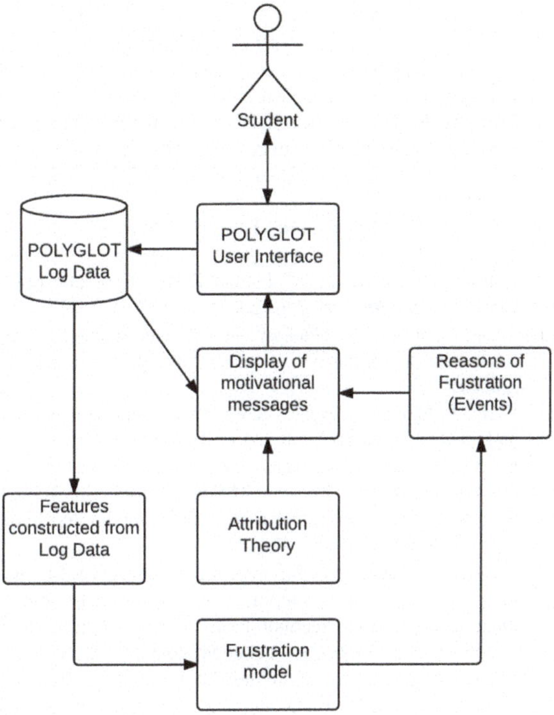

References

1. Baker, R., D'Mello, S.K., Rodrigo, M., Graesser, A.: Better to be frustrated than bored: the incidence, persistence, and impact of learners' cognitive-affective states during interactions with three different computer-based learning environments. Int. J. Hum.-Comput Stud. **68**(4), 223–241 (2010)
2. Calvo, R.A., D' Mello, S.K.: Affect detection: an interdisciplinary review of models, methods, and their applications. IEEE Trans. Affect. Comput. **1**, 18–37 (2010)
3. Conati, C., Maclaren, H.: Empirically building and evaluating a probabilistic model of user affect. User Model. User-Adap. Inter. **19**(3), 267–303 (2009)
4. Brawner, K., Goldberg, B.: Real-time monitoring of ECG and GSR signals during computer-based training. In: Intelligent Tutoring Systems, pp. 72–77, 2012
5. D'Mello, S.K., Craig, S., Gholson, B., Franklin, S.: Integrating affect sensors in an intelligent tutoring system. In: Affective Interactions: The Computer in the Affective Loop Workshop at ACM International Conference on Intelligent User Interfaces, pp. 7–13, 2005
6. Hussain, S., AlZoubi, O., Calvo, R.A., D'Mello, S.K.: Affect detection from multichannel physiology during learning sessions with autotutor. In: Proceedings of the 15th International Conference on Artificial Intelligence in Education (AIED 11), pp. 131–138, 2011
7. Sabourin, J., Mott, B., Lester, J.C.: Modeling learner affect with theoretically grounded dynamic bayesian networks. In: International Conference on Affective Computing and Intelligent Interaction, pp. 286–295, 2011
8. Kort, R., Reilly, R., Picard, R.W.: An affective model of interplay between emotions and learning: Reengineering educational pedagogy-building a learning companion. In: Advanced Learning Technologies—IEEE International Conference on Advanced Learning Technologies, pp. 43–46, 2001
9. Gee, J.P.: What video games have to teach us about learning and literacy. Comput. Entertain. **1**(1), 20–23, 2003
10. Kapoor, A., Burleson, W., Picard, R.W.: Automatic prediction of frustration. Int. J. Hum.-Comput. Stud. **65**, 724–736 (2007)
11. Lazar, J., Jones, A., Hackley, M., Shneiderman, B.: Severity and impact of computer user frustration: a comparison of student and workplace users. Interact. Comput. **18**(2), 187–207 (2006)
12. Klein, J., Moon, Y., Picard, R.W.: This computer responds to user frustration: theory, design, and results. Interact. Comput. **14**, 119–140 (2002)
13. Morgan, T., King, R.A., Weisz, J.R., Schopler, J.: Introduction to Psychology, 7th edn. McGraw-Hill Book Company, McGraw Hill Education, New York, USA (1986)
14. Spector, E.: Organizational frustration: a model and review of the literature. Pers. Psychol. **31**(4), 815–829 (1978)
15. Daish, S., Magidin de Kramer, R., O'Dwyer, L.M., Masters, J., Russell, M.: Impact of online professional development or teacher quality and student achievement in fifth grade mathematics. J. Res. Technol. Educ. **45**(1), 1–26 (2012)
16. McWhaw, K., Abrami, P.C.: Student goal orientation and interest: effects on students' use of self-regulated learning strategies. Contemp. Educ. Psychol. **26**(3), 311–329 (2001)
17. Jacob, A., Rockoff, J.E.: Organizing schools to improve student achievement: start times, grade configurations, and teacher assignments. Educ. Dig. **77**(8), 28–34 (2012)
18. Ewing, M.: Estimating the impact of relative expected grade on student evaluations of teachers. Econ. Educ. Rev. **31**(1), 141–154 (2012)
19. Graham, S., Weiner, B., Berliner, D., Calfee, R.: Theories and principles of motivation. In: Handbook of Educational Psychology, vol. 4, pp. 63–84 (1996)

Chapter 8
Overview of POLYGLOT Architecture and Implementation

Abstract This chapter presents an overview of the resulting social networking-based language learning system, called POLYGLOT, and an outline of its architecture. POLYGLOT is comprised of the following modules: Social Media User Interface module, Learning content module, Student model, Error diagnosis module, Win-Win Collaboration module, Frustration Recognition and Response module (Affective module), Learning style detection module and Adaptation model. An adequate number of screenshots of the operating system are also provided in the sections of this chapter in order to illustrate the intelligent and adaptive social e-learning process.

8.1 POLYGLOT Architecture

The research, presented in this book, involves a full development and implementation of the novel approach of a social and adaptive tutoring system incorporating machine learning techniques for the automatic detection of the student's learning style, error diagnosis mechanism and frustration management. Specifically, an innovative integrated e-learning environment for multiple language learning (English and French languages), which is called POLYGLOT, has been developed. The technology of Adaptive ITSs was taken into account for the system's design and development. Figure 8.1 depicts the model of the architecture of POLYGLOT. It consists of the following components:

- **Social Media User Interface module**: This module serves as the liaison between the learner and all the modules of the system. Its major characteristics are the user friendliness and the dynamic adaptation to each learner based on his/her needs and preferences [1, 2]. Towards this direction, this module should hold information concerning the learners' characteristics, needs and preferences along with good feedback about what's happening and whether the user's input is being successfully processed and mendable actions. Further characteristics include clarity, concision, responsiveness, consistency, familiarity, efficiency and forgiveness. Moreover, the user interface transfers the learning content to the users. Furthermore, the user interface of POLYGLOT consists of all the characteristics that social media have. Specifically, it has a wall on which all students can post their ideas, questions

C. Troussas and M. Virvou, *Advances in Social Networking-based Learning*, Intelligent Systems Reference Library 181, https://doi.org/10.1007/978-3-030-39130-0_8

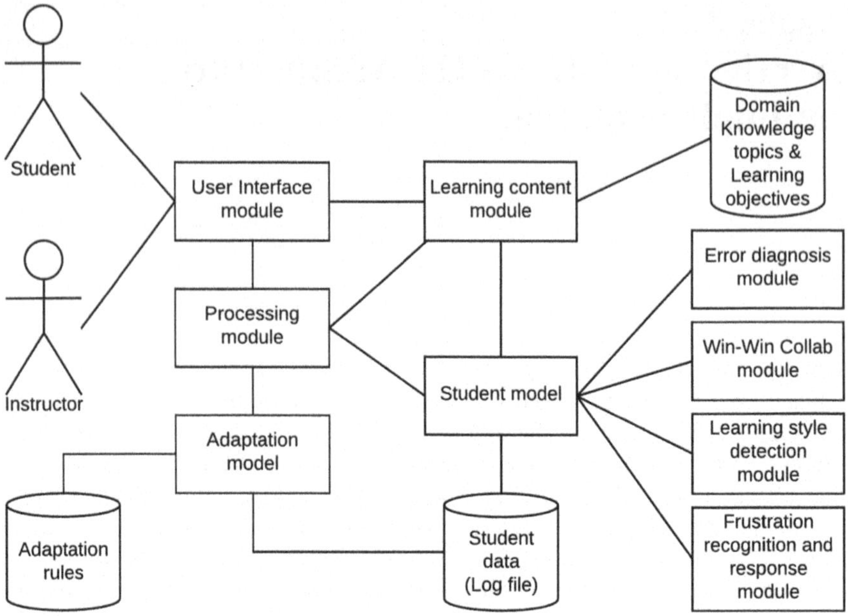

Fig. 8.1 The architecture of POLYGLOT

and they can interact with peers. Also, the students can tag their friends on the wall so that they can address to a specific person. Apart from that, students can send messages to other students or instructors in an instant or asynchronous way. Finally, the students can express their satisfaction or dissatisfaction concerning the exercises by pressing the "Like" or "Dislike" button respectively.

- **Learning content module:** The domain model contains knowledge pertaining to the subject matter. The system utilizes its domain knowledge to reason with and solve problems, or to answer questions posed by learners [3]. It is responsible to process the system domain knowledge to make inferences or solve problems [4]. Moreover, it provides explanations of problem solutions and gives alternative explanations of the same concept. Also, it answers arbitrary questions from the student and holds knowledge about common misconceptions and missing concepts. Finally, it incorporates the representation of the knowledge dependencies so that the status (namely if the learner has studied the material) and the difficulty level of the concepts can be analyzed.
- **Student model**: It is considered as the core component of an ITS paying special attention to student's cognitive and affective states and their evolution as the learning process advances [5]. As the learners work step-by-step through their exercise answering process, the ITS engages in a process called model tracing. Anytime the student model deviates from the domain model, the system identifies, or flags, that an error has occurred [6]. The student model is responsible to maintain information about the student's personal profile, knowledge, and current and

advancing skills. Furthermore, it stores information about the student's cognitive processes, learning preferences and/or past learning experiences. In this research, the aim is to model the cognitive states of each learner. Namely, the system has to be able to understand the learning state of each student and to recognize when a learner learns or not the learning content. To the direction of modeling the learner's knowledge, the overlay technique is used and it recognizes the progress that the student presents in the learning content. Given that the overlay model does not hold information about learners' errors and preferences, a stereotype model is used in addition. Moreover, a hybrid model of two different algorithms is used to interpret the nature of learners' errors. Furthermore, the system classifies learners into learning styles with the use of machine learning techniques.

- **Error diagnosis module:** It is responsible for diagnosing the misconceptions of students [7]. The error diagnosis module employs 2 different algorithms which can spot the type of error which is conducted by the student and the reason why s/he made it. POLYGLOT knowledge about how to solve an exercise correctly and in several faulty ways. The error diagnosis module uses a combination of buggy and overlay techniques to perform diagnosis of misconceptions. Buggy procedures are related to prerequisite grammatical concepts. Each one of these procedures is associated with a certain category of error. For example, a common mistake that students seem to make is the tense mistakes; namely, the student has neglected the rules of the proper use of tenses. The error diagnosis is performed by POLYGLOT in the Solving Exercises Mode (exercises where students must fill in the gap with the missing words). In multiple choice exercises error diagnosis is simple. For every erroneous answer that the student may select, there is an associated misconception. Therefore, depending on the erroneous selection that the student has made, a corresponding error message is presented, explaining the cause of the mistake. In the case of exercises where the student is asked to fill in the gap in a sentence, the error diagnosis becomes more sophisticated since in this case the student is allowed to be more creative than in multiple choice exercises. Hence, if the student's answer differs from the system's expectation then the system performs error diagnosis. Following, this module is further explained and described.
- **Win-Win Collaboration module**: Collaboration plays a crucial role in the tutoring process, since it creates a fertile ground to students to advance their knowledge level by cooperating with peers [8, 9]. It is responsible for recommending collaborations between learners with respect to either their learning state or the misconceptions that they conduct. By consulting win-win Collab module, the system provides advising to learners to collaborate with peers in such a way that both of them can reap the benefits of collaboration. The module offers two different approaches for collaboration. The first one is the win-win collaboration based on the already learnt language concepts and the second one is based on the types of misconceptions made by the student. For example, if a student is good at concept A but has poor knowledge on concept B, the system proposes him/her a collaboration with another learner who is complementary to the concepts. Also, under the same rationale, if a student is prone to conduct misconceptions of category A but s/he does not conduct

misconception of category B, the system proposes him/her collaboration with a student who conducts misconception of category B but not of category A.

- **Frustration Recognition and Response module (Affective module)**: The recognition of students' affective states especially when they are negative is very important for an educational aspect since the tutoring system can react and assist students when needed towards the unobstructed continuation of the educational process [10, 11]. It is responsible for providing personalized motivational messages to students in case of frustration. The system creates and displays messages to motivate the learners according to the reasons why the student is frustrated. The prime reason for frustration is goal failure. The possible reasons for goal failure are identified from the students' goal while they interact with the ITS.
- **Learning style detection module**: The determination of the students' learning style is important for creating a student-centric environment where each learner can learn in his/her own pace based on his/her needs, preferences and interests [12, 13]. This module involves the automatic detection of the student's learning style. POLYGLOT uses the Felder Silverman Learning Style Model and employs machine learning techniques in order to sophisticatedly select the right learning style of the student. This procedure does not involve traditional approaches for the detection of the learning style, such as questionnaires. In that way, the student saves a lot of time while POLYGLOT adapts the pace of tutoring to him/her based on his/her learning preferences.
- **Adaptation model**: It accepts information from the learning content and student model and makes choices about tutoring strategies and actions [14, 15]. At any point in the problem-solving process, the learner may request guidance on what to do next, relative to their current location in the model. In addition, the system recognizes when the learner has deviated from the production rules of the model and provides timely feedback for the learner, resulting in a shorter period of time to reach proficiency with the targeted skills. The adaptation model is aware of the progress of a learner and offers personalized tutoring and support. Also, it is responsible for providing tailored assessments to students based on their learning needs and preferences.

8.2 POLYGLOT Implementation

POLYGLOT is a web-based adaptive and intelligent system for foreign language learning, incorporating social features. POLYGLOT is programmed using the programming language JAVA. The following figures provide an overview of POLYGLOT.

Figures 8.2 and 8.3 show the log-in form and the registration form of POLYGLOT respectively. Figure 8.4 shows the start page of POLYGLOT. Figure 8.4 shows a start page of POLYGLOT. Figures 8.5 and 8.6 illustrate the statement of personal students' information and the preliminary test respectively. Through the preliminary test, POLYGLOT acquires information about the initial knowledge level of the student. Figure 8.7 illustrates the two different ways for detecting the learning style

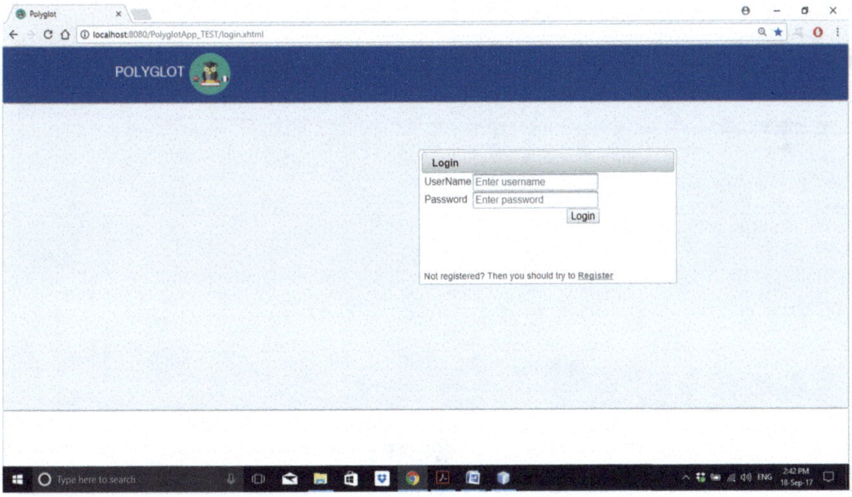

Fig. 8.2 Log-in form

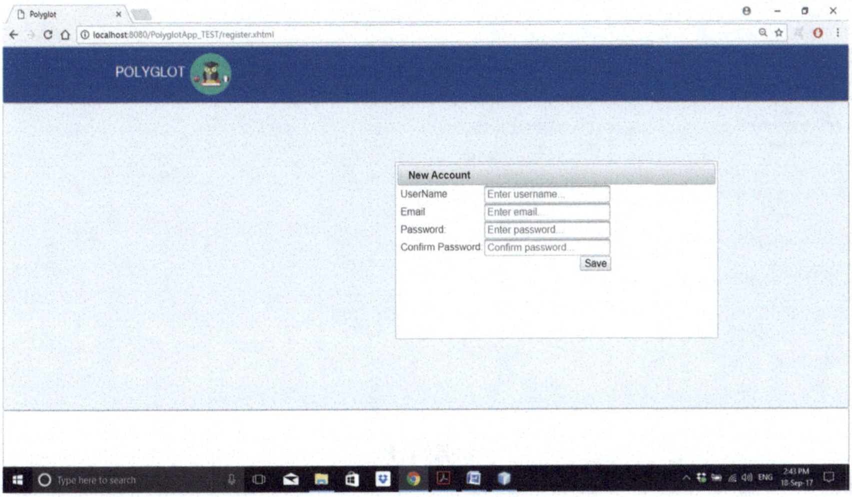

Fig. 8.3 Registration form

based on Felder and Silverman model; the first way is the automatic way, by simply pressing the corresponding button, while the second way is to answer the Felder and Silverman questionnaire, as shown in Fig. 8.8. Figures 8.9 and 8.10 show a sample of the learning content of the English and French languages respectively. Figures 8.11 and 8.12 illustrate a chapter test (multiple choice test) and the results of this test respectively. Figures 8.13 and 8.14 show the final test (fill-in the gaps questions) and its results respectively. Figure 8.15 illustrates the overall results, which each student

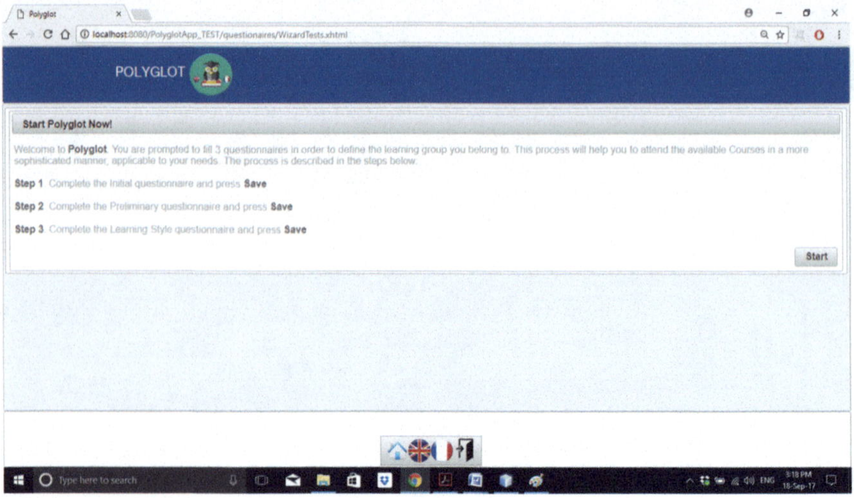

Fig. 8.4 Start page of POLYGLOT

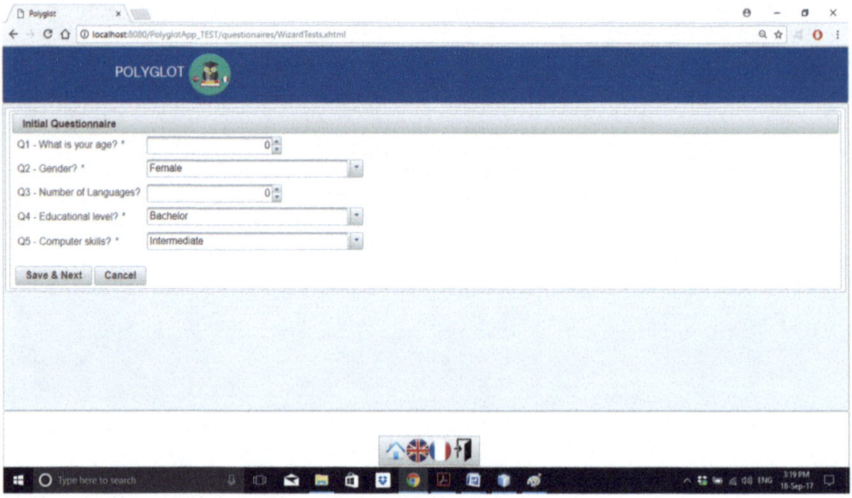

Fig. 8.5 Initialization of POLYGLOT's student model

can check along with charts that show graphically his/her progress in all the chapter tests and the final test. Figure 8.16 shows the wall on which each student can post along with the tagging activity, namely the capability of the student to tag the name of a classmate in order to address to him/her while posting on the wall. Figure 8.17 shows the notification message which notifies the student that a classmate tagged him/her. Figure 8.18 shows another way of communication between students or a student and the instructor through instant or asynchronous text messages. Figure 8.19

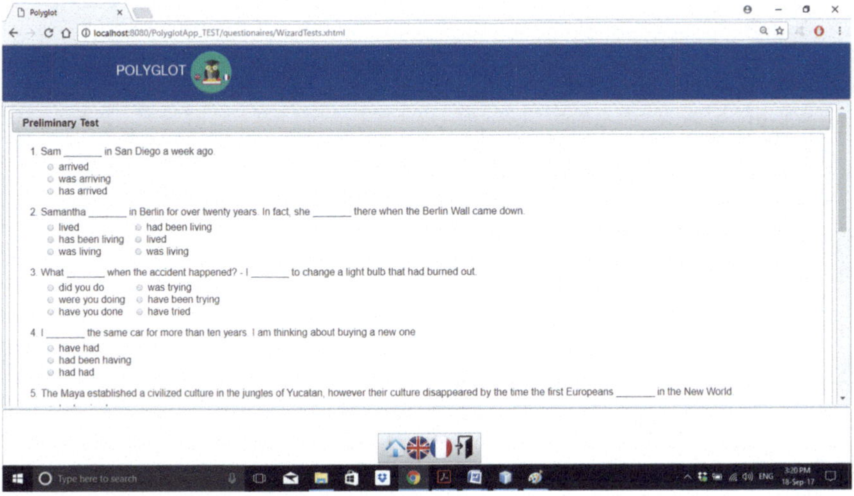

Fig. 8.6 Preliminary test

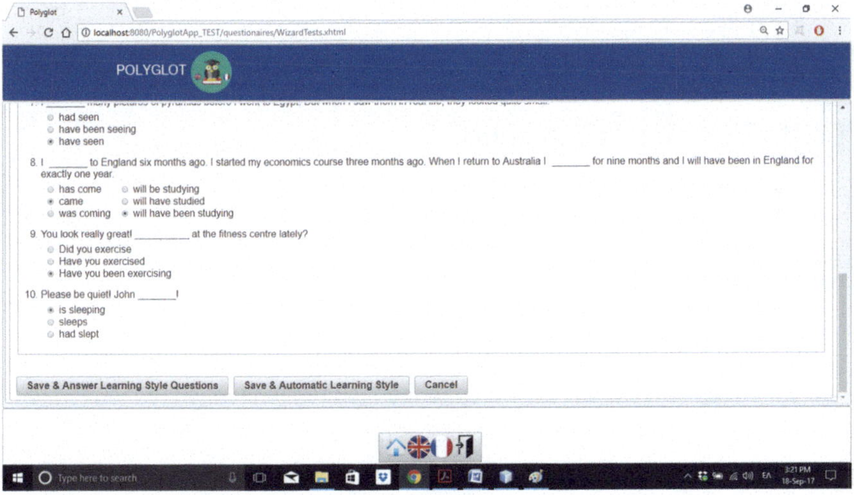

Fig. 8.7 Two ways of detecting students' learning styles

illustrates the declaration of a student's affective state, which may change after his/her interaction with POLYGLOT. Figure 8.20 shows a motivational message, which is delivered after the student's declaration of his/her affective state and before his/her interaction with POLYGLOT. After his/her interaction with POLYGLOT, namely taking part in an examination and liking/disliking the questions, the detection of frustration module is taking action and the motivational messages are tailored to his/her affective state. Figures 8.21 and 8.22 illustrate the two different ways of rec-

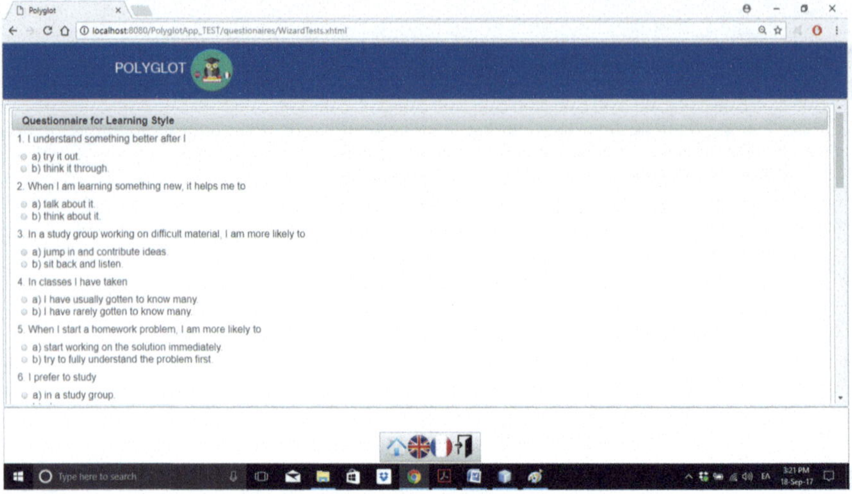

Fig. 8.8 Questionnaire to detect learning style

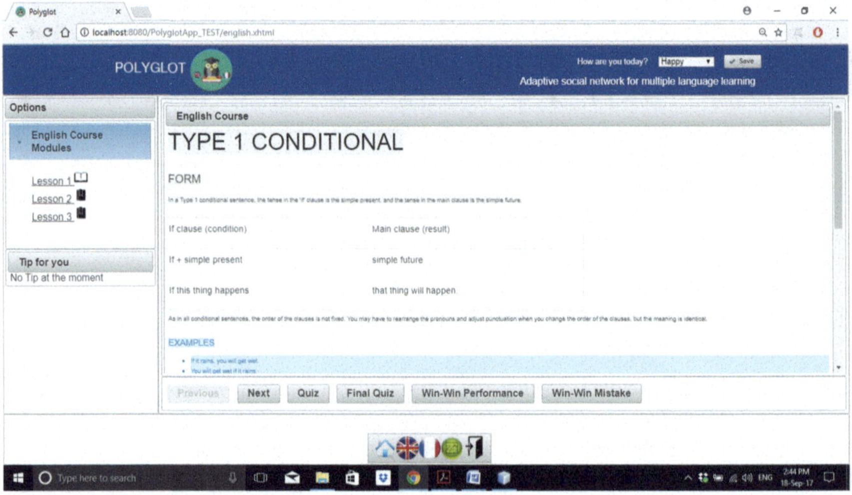

Fig. 8.9 Learning content in the English language

ommendation towards win-win collaboration concerning the student's knowledge level and type of conducted errors respectively. Figure 8.23 shows the first page of the authoring tool that the instructor can see. Figure 8.24 illustrates the authoring of the learning content of both foreign languages and also shows the authoring of the course quizzes. Finally, Fig. 8.25 shows information about the progress of each student along with a chart so that the instructor has a complete overview about the progress of the students.

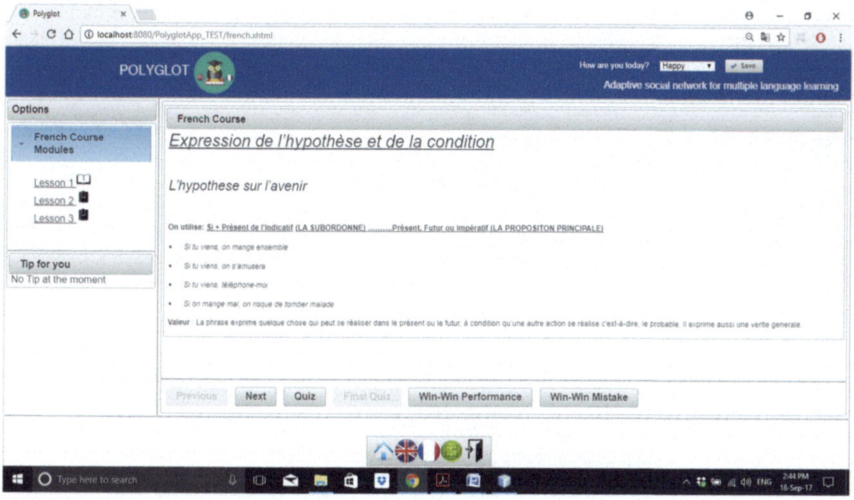

Fig. 8.10 Learning content in the French language

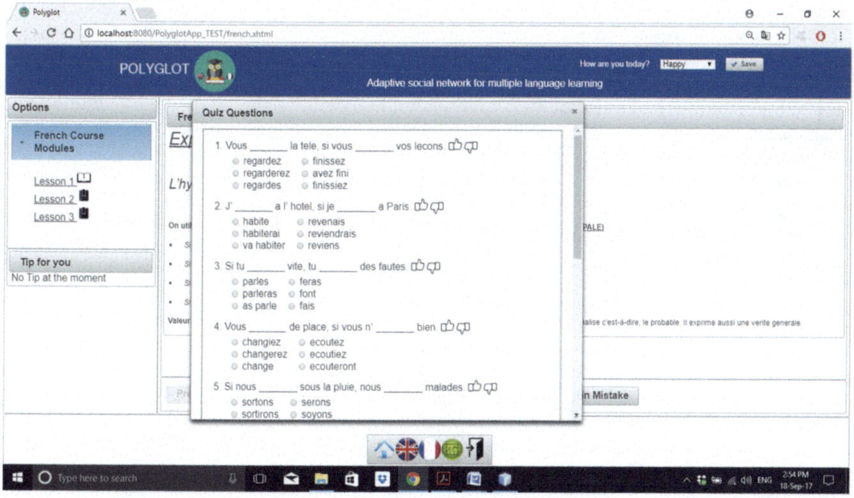

Fig. 8.11 Sample of chapter test (multiple choice exercise)

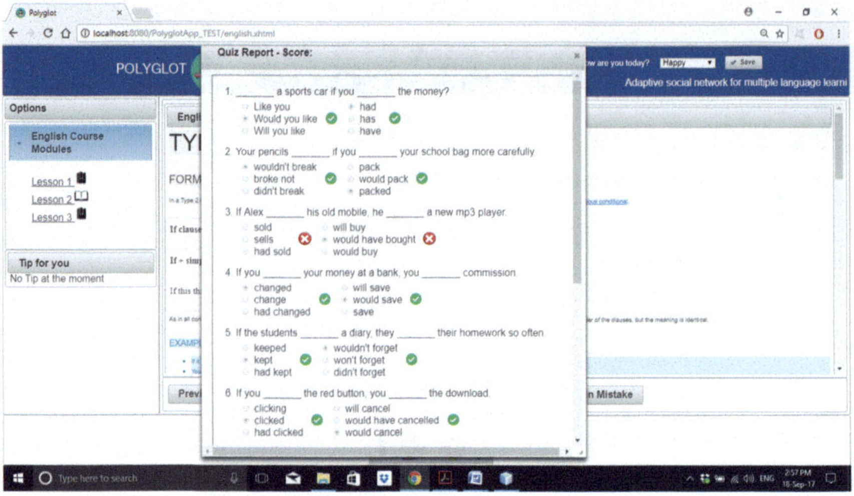

Fig. 8.12 Results of chapter test

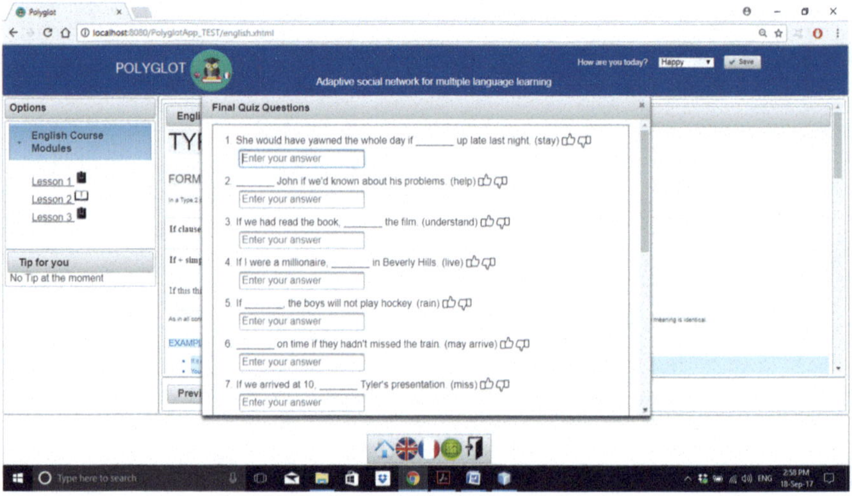

Fig. 8.13 Sample of the final test (fill-in the gaps exercise)

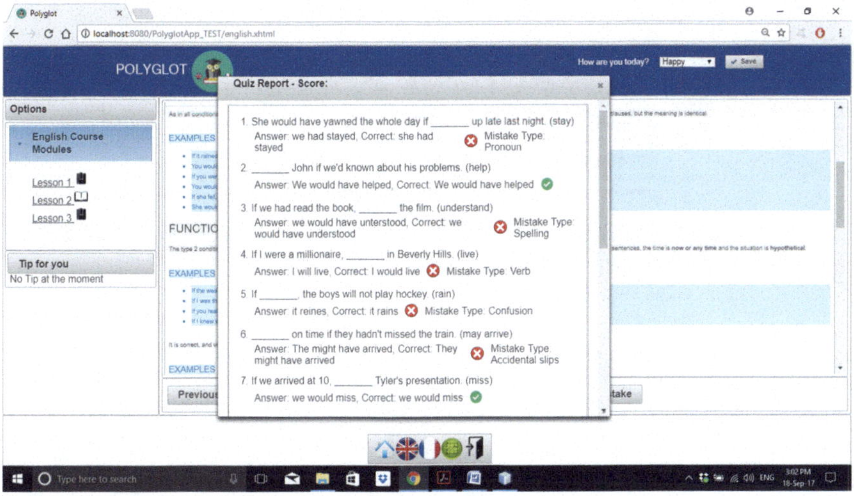

Fig. 8.14 Results of final test

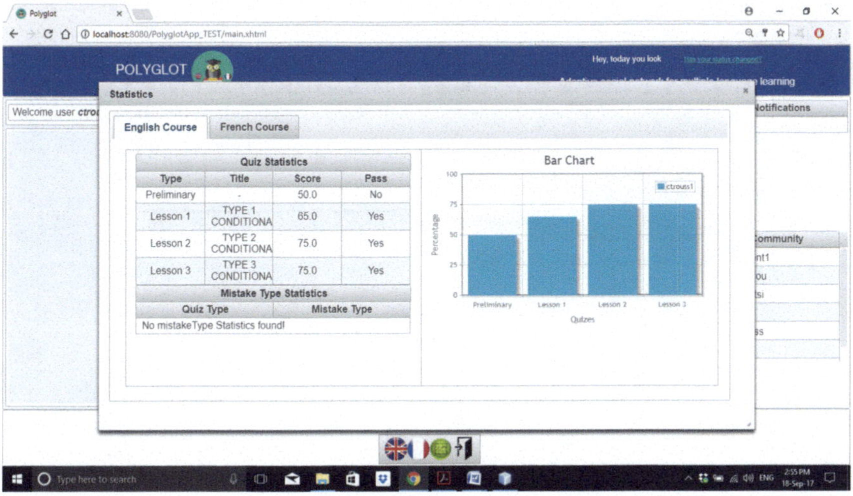

Fig. 8.15 Overall results of the student-progress

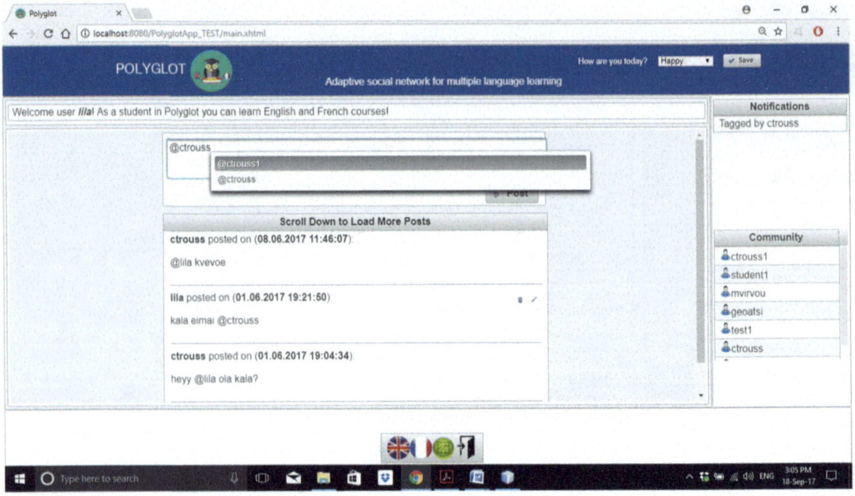

Fig. 8.16 Posting on wall and tagging a classmate

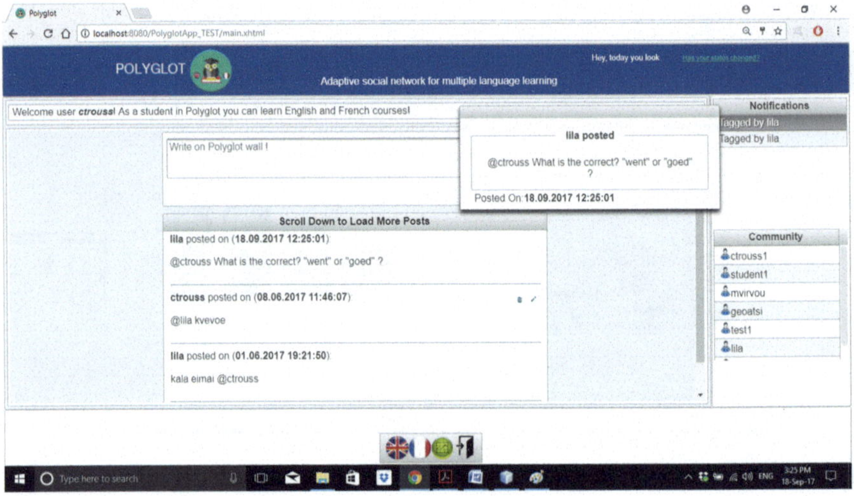

Fig. 8.17 Notification of tagging

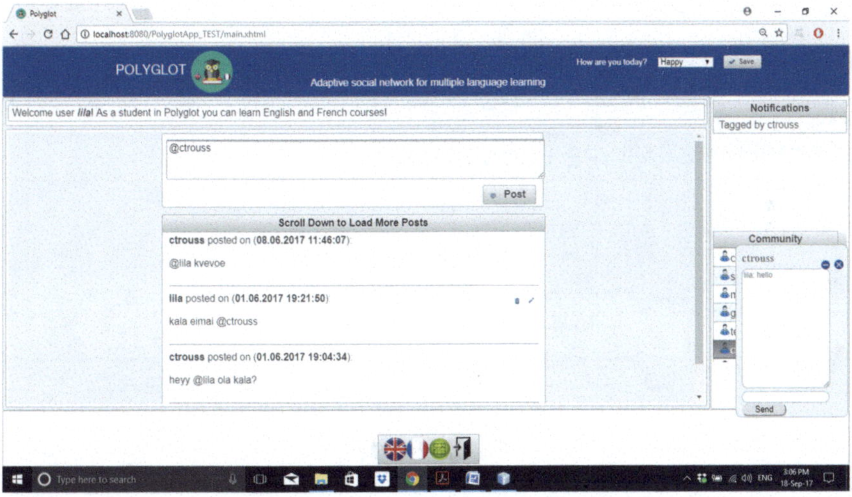

Fig. 8.18 Instant and asynchronous text messaging

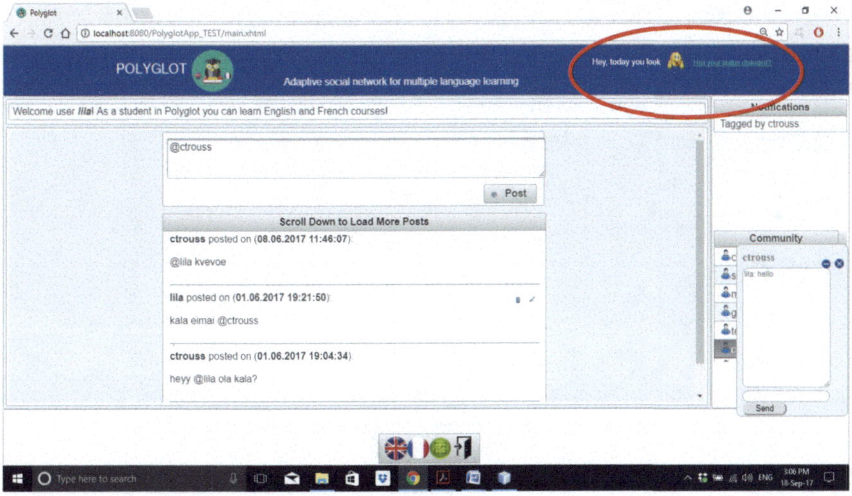

Fig. 8.19 Student's declaration of affective state

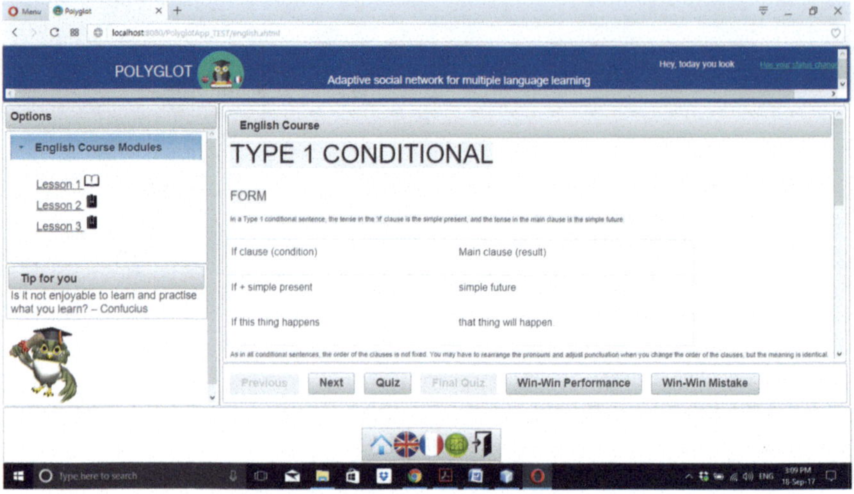

Fig. 8.20 Motivation message after the affective state declaration (and before taking a test)

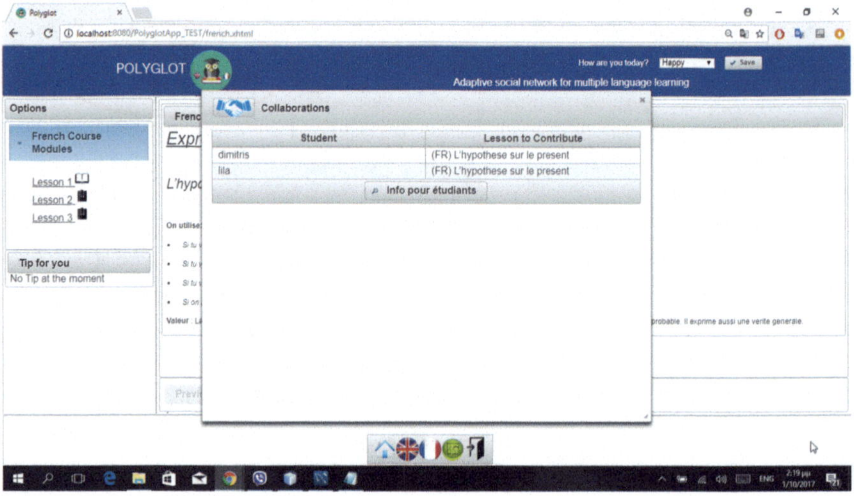

Fig. 8.21 Win-win collaboration based on knowledge level

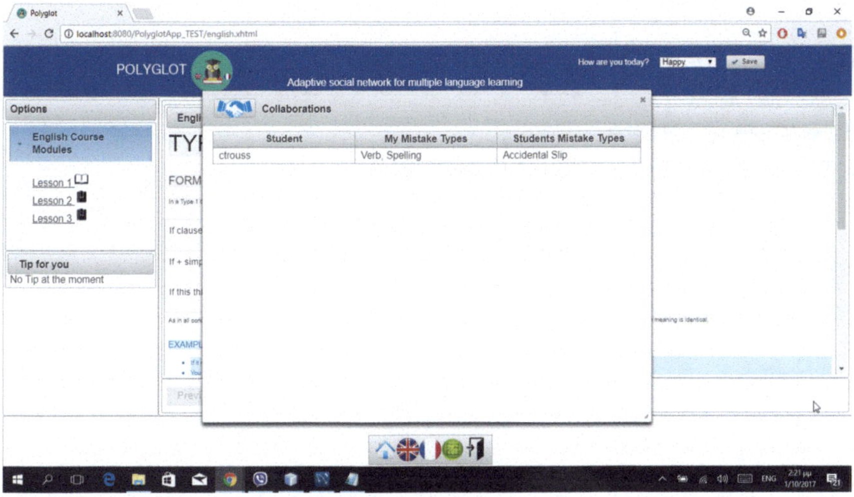

Fig. 8.22 Win-Win collaboration based on types of mistakes

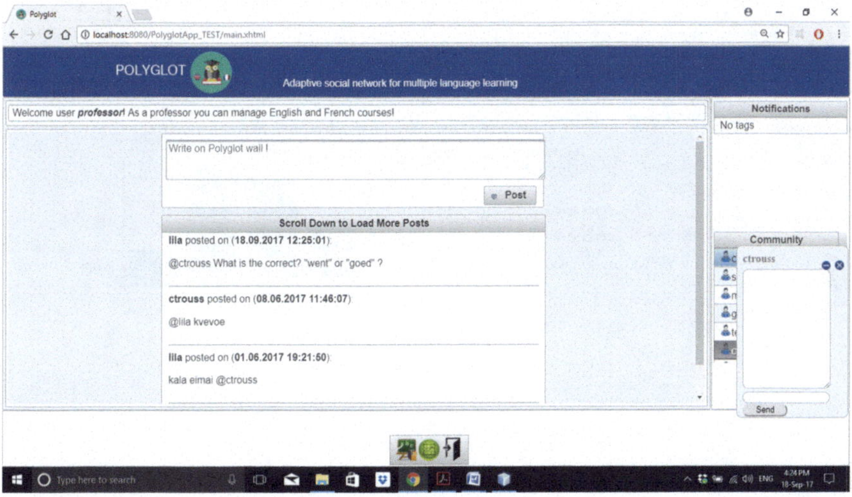

Fig. 8.23 Authoring tool-first page of instructor

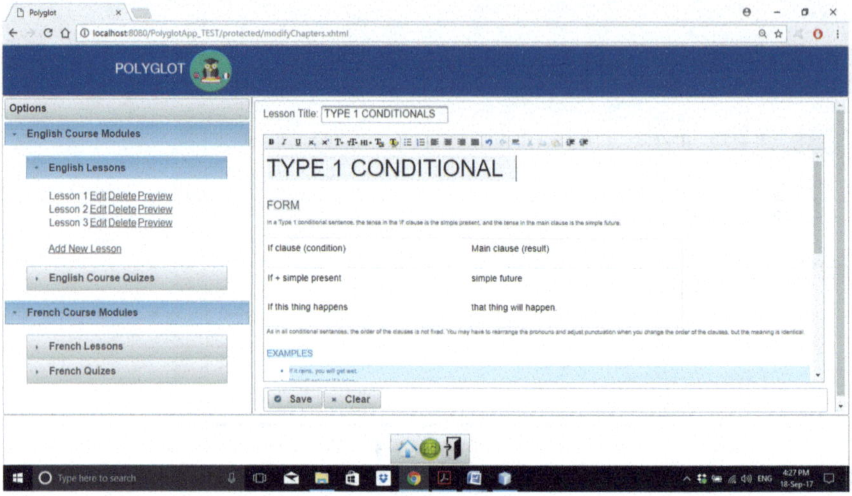

Fig. 8.24 Authoring of the learning content and the course tests

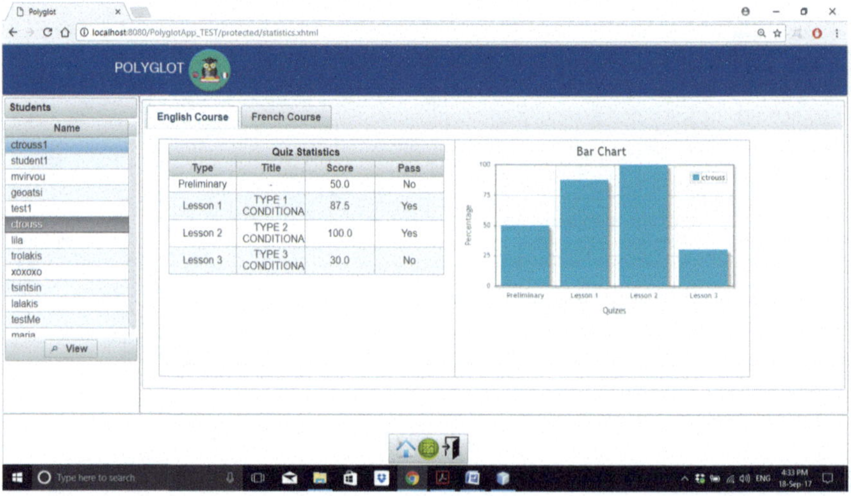

Fig. 8.25 Information and charts for students' progress

References

1. Jurdi, S., Garcia-Sanjuan, F., Nacher, V., Jaen, J.: Children's acceptance of a collaborative problem solving game based on physical versus digital learning spaces. Interact. Comput. **30**(3), 187–206 (2018)
2. Troussas, C., Krouska, A., Virvou, M.: Social interaction through a mobile instant messaging application using geographic location for blended collaborative learning. In: 2017 8th International Conference on Information, Intelligence, Systems & Applications (IISA), pp. 1–5. Larnaca (2017)

3. Troussas, C., Chrysafiadi, K., Virvou, M.: Machine learning and fuzzy logic techniques for personalized tutoring of foreign languages. In: Penstein Rosé C. et al. (eds.) Artificial Intelligence in Education. AIED 2018. Lecture Notes in Computer Science, 10948, pp. 385–362. Springer, Cham (2018)
4. Kim, D., Yoon, M., Jo, I.H., Branch, R.M.: Learning analytics to support self-regulated learning in asynchronous online courses: a case study at a women's university in South Korea. Comput. Educ. **127**, 233–251 (2018)
5. Troussas, C., Virvou, M., Espinosa, K.J.: Using visualization algorithms for discovering patterns in groups of users for tutoring multiple languages through social networking. J. Netw. **10**(12), 668–674 (2015)
6. Soledad, M., Grohs, J., Bhaduri, S., Doggett, J., Williams, J., Culver, S.: Leveraging institutional data to understand student perceptions of teaching in large engineering classes. In: 2017 IEEE Frontiers in Education Conference (FIE), pp. 1–8. Indianapolis, IN, (2017)
7. Troussas, C., Virvou, M., Vougiouklidou, A., Espinosa, K.J.: Automatic misconception diagnosis in multiple language learning over social networks. In: International Conference on Information, Intelligence, Systems & Applications (IISA), pp. 1–6. Piraeus (2013)
8. Erkens, M., Bodemer, D.: Improving collaborative learning: guiding knowledge exchange through the provision of information about learning partners and learning contents. Comput. Educ. **128**, 452–472 (2019)
9. Troussas, C., Virvou, M., Alepis, E.: Collaborative learning: group interaction in an intelligent mobile-assisted multiple language learning system. Inform. Educ. **13**(2), 279–292 (2014)
10. Troussas, C., Krouska, A., Virvou, M.: Reinforcement theory combined with a badge system to foster student's performance in e-learning environments. In: 2017 8th International Conference on Information, Intelligence, Systems & Applications (IISA), pp. 1–6. Larnaca (2017)
11. Troussas, C., Espinosa, K.J., Virvou, M.: Affect recognition through Facebook for effective group profiling towards personalized instruction. Inform. Educ. **15**(1), 147–161 (2016)
12. Krouska, A., Troussas, C., Virvou, M., Fragkakis, C.K.: Applying Skinnerian Conditioning for Shaping Skill Performance in Online Tutoring of Programming Languages. In: 2018 9th International Conference on Information, Intelligence, Systems and Applications (IISA), pp. 1–5. Zakynthos, Greece (2018)
13. Troussas, C., Krouska, A., Virvou, M., Sougela, E.: Using hierarchical modeling of thinking skills to lead students to higher order cognition and enhance social e-learning: In: 2018 9th International Conference on Information, Intelligence, Systems and Applications (IISA), pp. 1–5. Zakynthos, Greece (2018)
14. Jung, E., Kim, D., Yoon, M., Park, S., Oakley, B.: The influence of instructional design on learner control, sense of achievement, and perceived effectiveness in a supersize MOOC course. Comput. Educ. **128**, 377–388 (2019)
15. Krouska, A., Troussas, C., Virvou, M.: Computerized adaptive assessment using accumulative learning activities based on revised bloom's taxonomy. In: Virvou, M., Kumeno, F., Oikonomou, K. (eds.) Knowledge-Based Software Engineering: 2018. JCKBSE 2018. Smart Innovation, Systems and Technologies, vol. 108, pp. 252–258. Springer, Cham (2018)

Chapter 9
Evaluation Results for POLYGLOT and Discussion

Abstract In this chapter, the authors present the evaluation results and a discussion on them for the social networking-based language learning system, called POLYGLOT. All the approaches, mechanisms and models, presented in the previous chapters, are fully implemented and POLYGLOT is evaluated. The system was used by students of a private school of foreign languages in Athens, Greece in order to learn grammatical concepts in foreign languages. For the evaluation of all the modules of POLYGLOT, the Kirkpatrick's Four-Level Evaluation Model and the statistical hypothesis test were used. The results of the evaluation were very encouraging. They demonstrated that the system effectively adapts the learning process to the students' learning style while assisting them by diagnosing their misconceptions, recommending win-win collaborations, detecting their frustration and responding to this negative emotion.

9.1 Evaluation Process and Framework Used

Typically, systems evaluation is used to measure progress in achieving preset goals, ameliorate program development and provide useful feedback to instructors and learners. Posavac and Carey [9] observed that program evaluation is a "collection of methods, skills and sensitivities necessary to determine whether a human service is needed and likely to be used, whether the services are sufficiently intensive to meet the unmet needs identified, whether the service is offered as planned and whether the service actually helps people in need". In addition, McNamara [7] noted that improvement, in practice, implementation and reproduction is the goal of any high-quality program evaluation.

Evaluation can be valuable in different kind of software. It can either significantly assist in developing a concrete understanding of system's intended outcomes or give a clear perception of the system's efficiency. Moreover, systems evaluation does not concern solely the investigation of the relationships between expectations and outcomes; it has expanded to comprise more complex issues, such as effectiveness, efficiency, value and adequacy based on a systematic data collection and analysis

C. Troussas and M. Virvou, *Advances in Social Networking-based Learning*, Intelligent Systems Reference Library 181, https://doi.org/10.1007/978-3-030-39130-0_9

[10]. Nevertheless, system evaluations should produce a fertile ground for valid comparisons between similar programs [7].

There are many different types of evaluating measures depending on the objects or programs being assessed and the purpose of the evaluation [1]. The cornerstone of the evaluation is the manner in which information can be captured and used throughout the life of the program. McNamara [7] reports that the appropriateness of an evaluative measure has a direct correlation to the specific nature of information being sought. The judgment of the evaluation method is based on a specific methodology, a deep understanding of the information needed and knowledge from personal experiences and beliefs [1].

A system evaluation design depends on the information required in order to meet the objectives being set by the group seeking the evaluation [7]. As such, a focused evaluation that addresses the full set of objectives of a varied group of stakeholders will produce a qualitative result [1]. Furthermore, the overall goal to consider when selecting evaluation method is how to arrive at the most beneficial information to key stakeholders in the most realistic method.

Accordingly, evaluation is an inseparable part of tutoring systems. A teacher can do many things to collect information on the students' level of achievement. They include giving tests, assignments, oral questions, observation during the teaching-learning session, and portfolio. The activities are conducted not only to determine the students' grade but also to improve the quality of learning.

Learning evaluation should be conducted in a thorough and sustainable way, involving assessment on the learning process and outcomes. One of important factors that contribute to the achievement of educational objectives is the learning process itself.

On the other hand, evaluation and assessment (both on the learning process and on the outcomes in a continuous way) also play a role in encouraging the teaching staffs to improve the quality of learning process.

One of the main components in the education system is assessment. Assessment provides not only a description or information on the students' achievement or mastery of the learnt materials, but also a feedback to the educational program itself. Learning assessment is conducted as a part of decision-making process when it comes to the students' mastery of the materials after they are engaged in the teaching-learning process. In addition, learning assessment is also useful to figure out whether the learning strategy or approach is appropriate or not.

Accordingly, the educational system needs competent teaching staffs that are capable of not only teaching in a good way but also evaluating the learning outcome in an appropriate and effective way based on characteristics of the subject. As a part of the learning program, evaluation must be done in an optimum way. It should not rely merely upon the learning output, but also on the input, output, and quality of the learning process. In both educational sector and learning process, the role of information technology media should not be overlooked. The use of media is an element, which must be considered by the lecturers/teaching staffs in all of the learning activities. Accordingly, learning assessment should not rely merely upon the traditional tests.

Limitation of the traditional tests as the sole decision-making tool when it comes to the students' achievement is that it simply assesses the scientific knowledge. The assessment focuses only on the limited dimension of learning outcomes (knowledge and skills). It cannot be used to assess in-depth reasoning capability. In addition, it is not able to show the real competence of the students [8]. Another limitation of the traditional tests is that each question generally has a single, absolute answer. It does not focus on the process, but on the outcome; it neither reveals the students' thinking process nor measures all aspects of the teaching-learning process.

Mardapi [6] suggests that there are seven elements of learning evaluation. They are 1) focusing the evaluation, 2) designing the evaluation, 3) collecting information, 4) analyzing and interpreting, 5) reporting information, 6) managing evaluation, and 7) evaluating evaluation. The definition shows that in the early phases, an evaluator must first determine focuses and design of the evaluation.

The objective of evaluation is to obtain accurate and objective information on a program, which has been planned and implemented in the previous phases. The information may come from the process of program implementation, impacts/results, and efficiency. The results of evaluation determine whether the program is successful or not, whether it is going to be continued or stopped, and whether it is going to be used as a basis for the next program or not.

POLYGLOT was assessed using two different techniques. The one evaluation model that we use is the Kirkpatrick's model [4]. It defines four levels of evaluation:

- Level 1: Reaction: It examines how the students felt, and their personal reactions to the learning experience, for example:
  - did the trainees like and enjoy the training?
  - did they consider the training relevant?
  - was it a good use of their time?
  - did they like the venue, the style, timing, domestics, etc.?
  - level of participation
  - ease and comfort of experience
  - level of effort required to make the most of the learning
  - perceived practicability
- Level 2: Learning: This is the measurement of the increase in knowledge and intellectual capability from before to after the learning experience and concerns the following:
  - did the trainees learn what intended to be taught?
  - did the trainee experience what was intended for them to experience?
  - what is the extent of advancement or change in the trainees after the training, in the direction or area that was intended?
- Level 3: Behavior: This is the extent to which the trainees applied the learning and changed their behavior, and this can be immediately and several months after the training, depending on the situation and concerns the following:
  - did the trainees put their learning into effect when back on the job?

- were the relevant skills and knowledge used?
- was there noticeable and measurable change in the activity and performance of the trainees when back in their roles?
- was the change in behavior and new level of knowledge sustained?
- would the trainee be able to transfer their learning to another person?
- is the trainee aware of their change in behavior, knowledge, skill level?

- Level 4: Results: This is the effect on the business or environment resulting from the improved performance of the trainee. Measures would typically be business or organizational key performance indicators, such as: volumes, values, percentages, timescales, return on investment, and other quantifiable aspects of organizational performance.

9.1.1 Criteria

The definition of the evaluation should be defined initially. The proposed criteria are the following:

- Students' satisfaction about the e-learning system. Specifically, it concerns the degree of satisfaction in terms of the adaptation and effectiveness provided by the e-learning platform. Hence, the students' perspective towards the educational environment plays an important role.
- Students' performance. It concerns the performance of learners on the knowledge domain. Especially, it seeks to investigate the extent to which the learners gain knowledge on the taught concepts of the English and French languages.
- The changes that were caused on the individual state of the students. In other words, we want to assess the effect of the e-learning program on the behavior and thoughts of students about foreign language learning and distance learning.
- The results of the e-learning program to students' progress. It concerns the effects of the e-learning program to students' progress on their further studies.
- The validity of learning style detection, being done automatically by POLYGLOT for each student.
- The validity of recommendation for win-win collaboration between students seeking to cooperate with peers in a beneficial way for both parties.

9.1.2 Method

The method used for this evaluation coincides with the Kirkpatrick's model. Particularly, the assess of satisfaction coincides with the Kirkpatrick's evaluation of reaction level, the measurement of students' performance is similar to the Kirkpatrick's evaluation level of learning and the students' individual state/behavior and progress can be matched with the Kirkpatrick's evaluation levels of behavior and results, accordingly. Thus, the method of evaluation can be described as follows:

I. Assessing the learners' satisfaction about the e-learning environment. The level of satisfaction also involves the learner's satisfaction of the motivational messages that POLYGLOT delivered to them. For gathering this kind of information a questionnaire (Questionnaire A, Sect. 9.3) was used. The questions were close-ended based on Likert scale with five responses ranging from the low grade "Not at all" (1) to the high grade "Very much" (5). The questions were divided into two sections based on the type of information we were interested in. The questions of the first section were related to the effectiveness of the tutoring program. The second section was aimed at evaluating the adaptivity of the system.
II. Measuring the students' performance by conducting an experiment with an experimental group (the group of students which used the POLYGLOT environment) and a control group (the group of students which used a similar educational environment from which the student model was absent).
III. Assessing the changes on the students' state/behavior about language learning and e-learning. For gathering this kind of information a questionnaire (Questionnaire B, Sect. 9.3) was used. The questions were close-ended based on Likert scale with five responses ranging from the low grade "Not at all" (1) to the high grade "Very much" (5). The questions were divided into three sections based on the type of information we were interested in. The questions of the first section were related to the students' perception about language learning. The second section was aimed at evaluating the students' state towards e-learning. The third section included questions related to students' motivation to be involved in e-learning programs.
IV. Assessing the effects of the e-learning program on the students' progress concerning their further studies. For assessing this criterion, a questionnaire (Questionnaire C, Sect. 9.3) was used, which included five close-ended questions based on Likert scale with five responses ranging from the low grade "Not at all" (1) to the high grade "Very much" (5).
V. Assessing the validity of the detection of the learning style of the students being done in an automatic way at the first interaction of the student with POLYGLOT. More specifically, all the population taking part at the experiment (80 students) was asked to answer the Felder Silverman questionnaire in order to detect their learning style in a traditional way. After that, the results of the traditional learning style detection were compared to the results of the automatic learning style detection.
VI. Assessing the validity of recommendation for win-win collaboration, which support students' learning experience by proposing the proper classmate for cooperation. To this direction, based on the student models, POLYGLOT decides who is the proper student to propose to another student to work together so that the collaboration is advantageous and beneficial to both students involved. Hence, the only method to assess the validity of win-win collaboration

is to ask the learners' opinion about the collaboration and if it indeed helps them. Thus, a questionnaire (Questionnaire D, Sect. 9.3) with close-ended questions based on Likert scale with five responses ranging from "Not at all" (1) to "Very much" (5), was used.

9.1.3 Population

In total, the number of students that used POLYGLOT was 80. Apart from that, 20 users holding a degree in Informatics also used POLYGLOT. More specifically, POLYGLOT was used by a group of 40 students (group A) of a private school of foreign languages in Athens. After their participation in the training program, the learners completed the questionnaires A and D that are displayed in Sect. 9.3. After 6 months, the learners were asked to answer the questionnaires B and C (evaluation of behavior and the evaluation of results levels of the Kirkpatrick's model) that are displayed in Sect. 9.3. The answers of the above four questionnaires helped to assess students' satisfaction, the changes on students' state/behavior, the results on students' progress on their further studies and the validity of adaptation decision making.

Moreover, students' performance was measured and was compared with the performance of another group of 40 students (group B) of the same private school, which used a similar educational system from which all the mechanisms for adaptation and assistance were absent. Both systems had the same knowledge domain, which holds concepts in the English and French languages, but the second system delivers the concepts of the learning material in sequence without taking into account the students; learning style, error diagnosis, motivational messages recommendation for collaboration.

Learners of both groups had different ages, varying from 10 to 35, and backgrounds (Tables 9.1 and 9.2). Some of the students were primary or secondary school

Table 9.1 Distribution of students' ages and backgrounds

Ages	**10–14**	**15–18**	**19–25**	**26–30**	**31–35**
	28.36%	32.68%	14.24%	16.42%	8.30%
Background	**Primary/Secondary School students**		**University students**		**Working people**
	61.04%		16.21%		22.75%

Table 9.2 Distribution of students' knowledge of other languages

Language	English (%)	French (%)	English & French (%)
Group A	34.24	27.12	38.64
Group B	35.95	26.86	37.19

students, others were university students or people that already work. Furthermore, some of the students have computer skills. The number of students, which belong to either each age category or background category, is the same for both groups. The reason for this is the fact that the homogeneity of the experiment's samples simplifies the experiment's performing. The learners of both groups used the corresponding systems without attending any courses on language learning, over a period of six months.

9.2 Results

This section provides a detailed presentation of the evaluation results in terms of students' satisfaction, performance, individual state of learners, progress, validity of the detection of learning style and of win-win collaboration.

9.2.1 Satisfaction

As mentioned above, students' satisfaction coincides with the level 1 of the Kirkpatrick's model and as such it is very important for the evaluation of every learning environment. Based on the results of the questionnaires, the students' satisfaction about the adaptivity and effectiveness of POLYGLOT is high. Specifically, the students are very satisfied with the educational environment with the social characteristics and its contribution to the learning process. The results of the questionnaire are depicted in Fig. 9.1. This information is easy to collect, but does not tell enough about the learning success.

9.2.2 Performance

This is the evaluation given before, during, and after learning. The purpose of evaluating performance is to measure the degree to which learners have obtained knowledge based on their participation in the learning event. The evaluation conducted before learning determines the learners starting point. Each learner will have a different level of background knowledge prior to learning course material, so understanding where everyone stands to begin with allows for a more accurate measure. Evaluation during the learning event allows learners to self-evaluate, and measure their own progress. It also gives facilitators a sense of how well learners are doing in relation to the learning objectives. The evaluation at the end of the learning event is also referred to as a summative evaluation, and it is done individually. According to Hamtini [3],

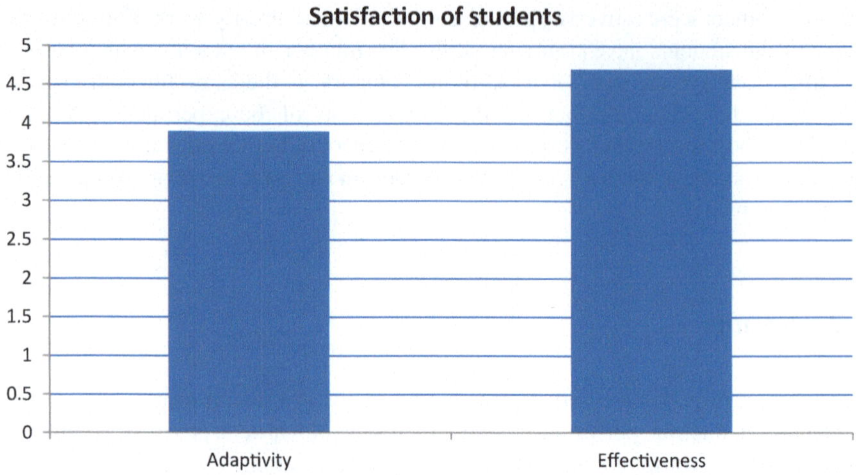

Fig. 9.1 Students' satisfaction

LaMotte [5] and Galloway [2], the most appropriate method of evaluation is to conduct pre-tests and post-tests. For this reason, Student's t-Tests were chosen to conduct this evaluation.

In the evaluation study, 80 students from different classrooms participated. As mentioned above, the students were all from a private school of foreign languages. The school, that was chosen, is located in Athens, the capital city of the country. Hence, it can be seen as a representative sample, since it adequately replicates the larger statistical population in terms of students' characteristics. School teachers also provided very valuable help in the whole evaluation study since they also participated both in the use of the ITS from the students and also provided assistance to their students while they interacted with the educational platform. The first group evaluated POLYGLOT, while the second group evaluated an ITS offering the same learning material and tests but without the same user interface all the modules of POLYGLOT. This division was very crucial in order to compare the performance of students using POLYGLOT in comparison with a simple e-learning platform. As a result, both groups had given a brief presentation on how to use the educational platform. Consequently, each group had the appropriate knowledge and enough time (6 months) to spend interacting with POLYGLOT. After the completion of their interaction (group A with POLYGLOT and group B with simple e-learning platform), all students were given questionnaires to complete with guidance from the evaluators and also their teachers.

The evaluation study was conducted with the use of self-supplemented scale questionnaires incorporating closed questions for the students. For this research, the Questionnaire C is used.

Table 9.3 Statistical significance in a student's t-Test for question 1

t-Test: Two-sample assuming equal variances		
	Variable 1	*Variable 2*
Mean	2.9	4
Variance	1.476923077	0.666666667
Observations	40	40
Pooled Variance	1.07179487179487	
Hypothesized Mean Difference	0	
Degrees of freedom	78	
t	−4.751730987	
P(T < = t) one-tail	4.52	
t Critical one-tail	1.664624645	
P(T < = t) two-tail	9.03057	
t Critical two-tail	1.990847036	

It was observed that students became familiar easily and very quickly with the educational software, its features and its functionalities. Their interest was undiminished during the whole 6-month period of their interaction with the educational application.

Finally, Tables 9.3, 9.4, 9.5, 9.6 and 9.7 illustrate the statistical significance of the questions 1–5 respectively (Questionnaire C, Sect. 9.3). Assuming the null hypothesis, the probability of this result is 0. As such, for the null hypothesis "There is no difference between the two groups of students", the t-Test rejects the hypothesis for all the questions. The absolute value of the calculated t exceeds the critical value, so

Table 9.4 Statistical significance in a student's t-Test for question 2

t-Test: Two-sample assuming equal variances		
	Variable 1	*Variable 2*
Mean	2.85	4.275
Variance	1.515384615	0.51217949
Observations	40	40
Pooled Variance	1.01378205128205	
Hypothesized Mean Difference	0	
Degrees of freedom	78	
t	−6.32932743	
P(T < = t) one-tail	7.20749	
t Critical one-tail	1.664624645	
P(T < = t) two-tail	1.4415	
t Critical two-tail	1.990847036	

Table 9.5 Statistical significance in a student's t-Test for question 3

t-Test: Two-sample assuming equal variances		
	Variable 1	*Variable 2*
Mean	2.8	4.25
Variance	1.497436	0.602564
Observations	40	40
Pooled Variance	1.05	
Hypothesized Mean Difference	0	
Degrees of freedom	78	
t	−6.32832	
P(T < = t) one-tail	7.24	
t Critical one-tail	1.664625	
P(T < = t) two-tail	1.45	
t Critical two-tail	1.990847	

Table 9.6 Statistical significance in a student's t-Test for question 4

t-Test: Two-sample assuming equal variances		
	Variable 1	*Variable 2*
Mean	2.325	4.325
Variance	1.250641	0.430128
Observations	40	40
Pooled Variance	0.840384615384615	
Hypothesized Mean Difference	0	
Degrees of freedom	78	
t	−9.75677	
P(T < = t) one-tail	1.85	
t Critical one-tail	1.664625	
P(T < = t) two-tail	3.71	
t Critical two-tail	1.990847	

the means are significantly different. Hence, it is concluded that the tutoring system has a statistically significant effect on performance.

It was expected that younger students with an inherent tend towards new technology would welcome e-learning learning with social characteristics adapted to their needs, supporting their learning. The findings of this preliminary study are rewarding the authors' attempts towards moving education to the fast growing field of intelligent tutoring systems incorporating social features and adaptivity. Analyzing the results of the evaluation study there is considerable evidence that this new technology is quite welcome from young learners and could be incorporated in schools supporting the educational process. The above tables illustrate that the performance of students

Table 9.7 Statistical significance in a student's t-Test for question 5

t-Test: Two-sample assuming equal variances		
	Variable 1	*Variable 2*
Mean	2.375	4.425
Variance	1.112179	0.353205
Observations	40	40
Pooled Variance	0.732692307692307	
Hypothesized Mean Difference	0	
Degrees of freedom	78	
t	−10.7105	
P(T < = t) one-tail	2.77	
t Critical one-tail	1.664625	
P(T < = t) two-tail	5.54	
t Critical two-tail	1.990847	

using POLYGLOT was exceptionally high and as such POLYGLOT serves as a great tool for learning.

9.2.3 Individual State of Learners

The individual state of the learners along with their behavior has significantly changed in a more positive level. The interaction with POLYGLOT notably ameliorated the students' perspective and opinion towards the language learning and e-learning. The results showed that the students are very keen on using an e-learning platform for learning foreign languages. This fact is attested by the teachers of the private school of foreign languages who assured that the students were very interested in using POLYGLOT for learning the taught concepts. In order to enhance the accuracy of the results, students were divided in two distinct categories. The first category includes students who are prone to foreign language learning, while the second category includes students with no foreign language knowledge. It should be clarified that the proneness to foreign language learning means that the students are very keen on learning foreign languages or they are novice, intermediate or expert in one or more foreign languages. The reason why students were categorized as mentioned is because of the fact that the changes in the state of students who are prone to foreign language learning may be less important. Furthermore, it should be noted that POLYGLOT takes into consideration the previous level of knowledge in the use of computers. The students having been involved in the experiment had a high level of knowledge in the use of computers. As such, they do not meet any obstacle in using POLYGLOT and they focus on the instruction issues. The questionnaire B that was answered by the learners and the mean of students' answers are displayed later in this section. The results of the questionnaire are depicted in Figs. 9.2 and 9.3.

The results show that the students' state towards foreign language learning (specifically English and French languages) and e-learning, who are not prone to foreign language learning or who had no previous knowledge on foreign languages, was improved by 81.1 and 78.3% respectively. While their willingness to be engaged in e-learning programs, was increased by 76.2%. Similarly, the state of the learners, who are prone to language learning and namely who have been involved in the learning

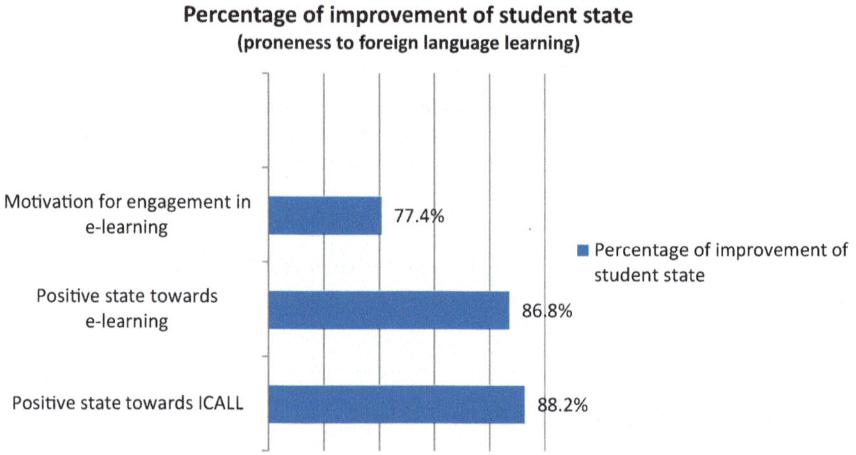

Fig. 9.2 Changes on individual state of students with no previous knowledge on foreign languages (no proneness to language learning)

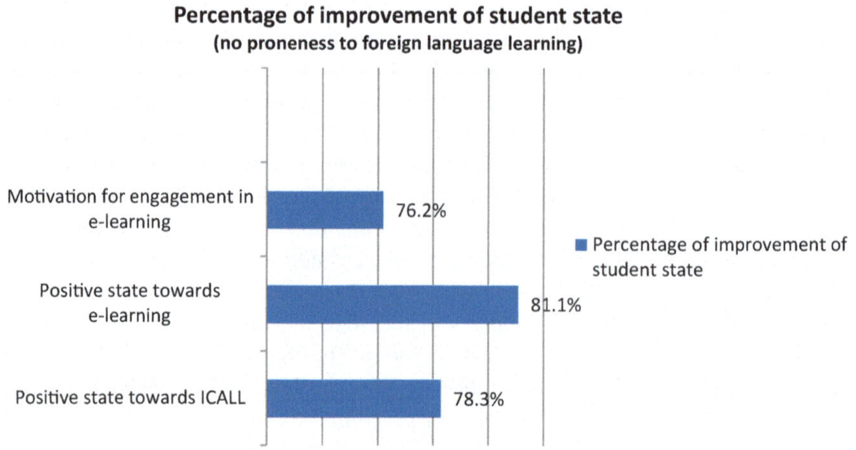

Fig. 9.3 Changes on individual state of students with previous knowledge on foreign languages (proneness to language learning)

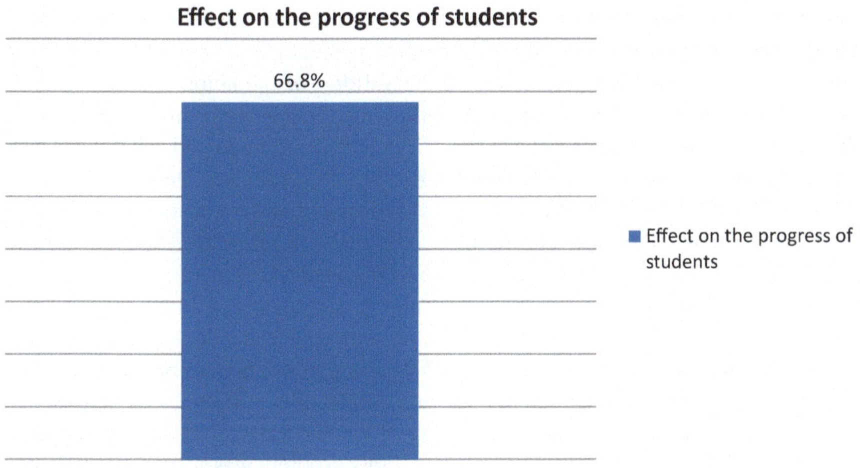

Fig. 9.4 Results on learners' progress

of at least one foreign language, towards foreign language learning and e-learning was improved by 86.8 and 88.2% respectively. Also, their motivation to be involved in e-learning programs was increased by 74.4%.

9.2.4 Students' Progress

The results of the e-learning program to the learners' progress on their further studies are satisfactory. The results of the questionnaire reveal that the e-learning program helped the users. The questionnaire C that was answered by the learners is displayed later in this section, while the results are depicted in Fig. 9.4. The teachers of the students in the private school of foreign languages along with the grades of the tests (on the concepts being taught in POLYGLOT) which were delivered to students after the period of using POLYGLOT can confirm the aforementioned results.

9.2.5 Validity of the Detection of the Students' Learning Style

The detection of the students' learning style seems to be very satisfactorily valid. According to the results, POLYGLOT's automatic detection coincides with the traditional discovery (discovery based on the Felder Silverman questionnaire) of the learning style, giving the impressively high percentage of 95%. More specifically, after their interaction with POLYGLOT, students were asked to fill in the Felder-Silverman questionnaire in order to check if the automatic detection of their learning style coincides with the results of the questionnaire. After the students' interaction with POLYGLOT, they were also asked to answer if they are satisfied with the learning style which POLYGLOT detected for them; the percentage of students' satisfaction was again 95%. The high percentage of the validity of the automatic detection of the learning style was almost expected. Following, the reason for this expectation is clarified. As mentioned above, the automatic detection is conducted using the k-NN, which is a supervised machine learning algorithm. A supervised learning algorithm analyzes the training data and produces an inferred function, which can be used for mapping new examples. The optimal scenario will allow for the algorithm to correctly determine the class labels for unseen instances. Hence, this requires the learning algorithm to generalize from the training data to unseen situations in a "reasonable" way. To this direction, the algorithm was rendered able to learn to predict a certain target output. To achieve this, k-NN was given 100 training examples that demonstrate the intended relation of input and output values. Then it was supposed to approximate the correct output, even for examples that have not been shown during training. With several additional assumptions, this problem was solved exactly since unseen situations might have an arbitrary output value.

9.2.6 *Validity of Win-Win Collaboration*

The results of the validity of the recommendation for win-win collaboration were positive (Questionnaire D, Sect. 9.3). According to the results, 85% of the students liked the experience by stating that they had a fruitful collaboration with the right classmate. Furthermore, 90% of the students took assistance from this process by collaborating with a classmate who has complementary knowledge level or conducts different type of mistakes. The above percentages are sufficiently satisfactory to be able to lead to the conclusion that the recommendation for win-win collaboration is proper and supports the tutoring process.

9.3 Questionnaires

Following, the questionnaires, which have been used for the evaluation, are presented.

Questionnaire A

	Questions
Effectiveness	Does the educational software meet your expectations?
	Does the educational software help you understanding the rationale of learning foreign languages?
	Do you think that this educational software is useful as an educational "tool"?
	Do you think that the use of this educational software is a waste of time?

(continued)

(continued)

	Questions
	After the end of the educational process, do you feel that you have assimilated all the subjects that you are taught?
Adaptivity	Does the program correspond to your knowledge level each time?
	Does the program correspond to your educational needs level each time?
	How time do you spend on issues that you already known?
	Does the test adapt to your educational needs?
	Does the learning style which POLYGLOT picked for you match to your needs?
	Do the motivational messages assist you on language learning?

Questionnaire B

	Questions
State on foreign language learning	Does the educational software affect positively your perception about foreign language learning?
	Does the educational software draw your interest on foreign language learning?
	Does the educational software motivate you to be involved in foreign language learning?
State on e-learning	Does the educational software help you to understand the subject of computers in education?
	Does the educational software affect positively your perception about distance learning?
Engagement in e-learning	Does the educational software motivate you to deal with distance education?
	Does the educational software motivate you to join other e-learning programs?

Questionnaire C

Questions
Does the educational software help you understanding better concepts on foreign language learning?
Does the educational software help you to learn other foreign languages?
Does the educational software help you in your studies?
Does the educational software help you understanding other lessons related to language learning?
Does the educational software help you in the elaboration of tasks and activities considering your studies?

Questionnaire D

Questions
Do you think that the person that POLYGLOT recommended to you for collaboration was the right one in terms of helping each other?
Is the collaboration with your classmate (proposed by POLYGLOT) fruitful?
Do you believe that you take and receive assistance from the proposed classmates having complementary knowledge level or type of misconceptions?

References

1. Fitzpatrick, J.L., Saners, J.R., Worthen, B.R.: Program Evaluation: Alternatives, Approaches, and Practical Guidelines, 3rd edn. Pearson Education Inc., Boston (2004)
2. Galloway, L.: Evaluating distance delivery and e-learning is Kirkpatrick's model relevant? Perform. Improv. **44**(4), 21–27 (2005)
3. Hamtini, T.: Evaluating e-learning programs: an adaptation of Kirkpatrick's model to accommodate e-learning environments. J. Comput. Sci. **4**(8), 693–698 (2008)
4. Kirkpatrick, D.L.: Techniques for evaluating training programs. Train. Dev. J. **33**(6), 78–92 (1979)
5. LaMotte, A.: Measure the effectiveness of your e-learning course with Kirkpatrick's 4 levels of evaluation. E-learning Heroes (2015). https://community.articulate.com/articles/kirkpatrick-model-of-training-evaluation-for-e-learning
6. Mardapi, D.: Curriculum and school educational evaluation system optimization. National Seminar on Competency-Based Curriculum. Ahmad Dahlan University, Yogyakarta (2004)
7. McNamara, T.: Language Testing. Oxford University Press (2000)
8. Mokhtari, K., Yellin, D., Bull, K., Montgomery, D.: Portfolio assessment in teacher education: impact on preserve teachers' knowledge and attitudes. J. Teach. Educ. **47**(4), 245–252 (1996)
9. Posavac, E., Carey, R.: Program Evaluation: Methods and Case Studies, 7th edn. Pearson Prentice Hall, Upper Saddle River, NJ (2007)
10. Rossi, P., Freeman, H., Lipsey, M.: Evaluation: A Systematic Approach, 6th Edn. Sage, Revised Edition

Chapter 10
Conclusions

Abstract In the last chapter of this book, the authors present their conclusions, derived from their recent research studies over the advances of e-learning arising from the utilization of social networks in this field. Indeed, the utilization of social networks in the field of e-learning brings considerable changes in the domain of knowledge acquisition since the features and capabilities of social networks create a new era in technology-aided instruction. Thus, a new research area is created, that of intelligent social networking-based learning systems. More specifically, the authors' attempts were targeted to the domains of e-learning, social networks, artificial intelligence, affective computing and cognitive and learning theories. Their proposed architectures, mechanisms, models, techniques and methodologies have concluded in analyzing, designing and implementing actual software, which was also fully evaluated by real end users. This book's last discussion reveals a fulfilment in the authors' efforts in the related scientific fields, as well as their suggestions and challenges for future work in intelligent social networking-based learning software.

10.1 Conclusions and Discussion

Intelligent and adaptive techniques can be employed in e-learning systems with great results in supporting personalized learning [1, 3]. The fast pace of everyday life in terms of modern technology utilization renders social interaction an indispensable tool for e-learning systems [4, 2]. The objective of this research was to create a novel social e-learning system which provides adaptive and personalized instruction to students. The developed system incorporates social characteristics and particularly posting on a wall, tagging a classmate, instant, declaring the affective state, liking/disliking of the exercises and instant and asynchronous text messaging. As such, the learning process takes place in an already familiar interface given that nowadays people spend a lot of their spare time in social networking sites, such as Facebook, and are very aware of this technology. Furthermore, POLYGLOT employs machine learning techniques, namely the k-nearest neighbors algorithm, in order to automatically define the learning style of the student based on the Felder-Silverman Learning Style Model. As such, the user does not need to answer a great deal of

C. Troussas and M. Virvou, *Advances in Social Networking-based Learning*, Intelligent Systems Reference Library 181, https://doi.org/10.1007/978-3-030-39130-0_10

the questions proposed by the aforementioned model. Thereby, POLYGLOT can infer about the way with which the student prefers to process information (active and reflective learners) and the student progress towards understanding (sequential and global learners). The learning style of the student adapts the program on the student based on his/her preferences and needs. Therefore, the system allows each individual learner to complete the e-learning course at a friendlier interface that takes into consideration the individuality of the learners in terms of the way and pace of learning. In this way, the system helps learners to save time and effort during the learning process.

Moreover, POLYGLOT supports win-win collaboration. More specifically, the algorithmic techniques that have been used serve as a recommendation tool to students and assist them concerning the right classmate to choose for collaboration. The system incorporates two different approaches for collaboration. The first one is the win-win collaboration based on the already learnt language concepts. The second approach concerns the types of misconception that the user made. For example, if a student is good at concept A but has poor knowledge on concept B, the system proposes him/her a collaboration with another learner who is complementary to the concepts. Also, under the same rationale, if a student is prone to conduct misconceptions of category A but s/he does not conduct misconception of category B, the system proposes him/her collaboration with a student who conducts misconception of category B but not of category A. As such, based on two significant characteristic, namely the gained knowledge on taught concepts and the type of students' misconceptions, the system recommends collaboration between classmates and they will both learn from each other. To this direction, the system provides advising to learners to collaborate with peers in such a way that both of them can reap the benefits of collaboration and learn while collaborating.

Moreover, POLYGLOT employs an error diagnosis module in order to successfully recognize the categories of errors that students make. The types of misconceptions that are diagnosed by the system are the accidental slips, pronoun mistakes, spelling mistakes, verb tense mistakes, language transfer interference. For this reason, two algorithmic approaches are incorporated. The first technique is the approximate string matching which finds string similarities by matching a student's given "exact" wrong answer with the systems correct stored answer. This technique is responsible for finding strings that match a pattern approximately. The problem of approximate string matching is typically divided into two sub-problems: finding approximate substring matches inside a given string and finding dictionary strings that match the pattern approximately. If string matching occurs in a high percentage, POLYGLOT decides whether the mistake lies among the categories of accidental slips, pronoun mistakes, spelling mistakes or verb mistakes. Correspondingly, using the second technique of string meaning similarity, POLYGLOT also tries to find meaning similarities between the given and the correct answer by translating these two answers to the system's available supported languages, namely the English and French languages. As such, the type of Language Transfer Inference mistake can be detected and diagnosed. Towards this direction, POLYGLOT can perform misconception detection and diagnosis so that POLYGLOT holds this information and assists the student in the tutoring process. Furthermore, POLYGLOT employs techniques for tailored assessment to students. These techniques involve multiple-criteria decision analysis for delivering exercises to students in an individualized way.

Also, one main innovation of the implemented system is the provision of personalized motivational messages to students in case of frustration. The system creates and displays messages to motivate the learners according to the reasons why the student is frustrated. The prime reason for frustration is goal failure. The possible reasons for goal failure are identified from the students' goal while they interact with the ITS. Upon the first interaction of the student with POLYGLOT, s/he can state his/her affective state. This adheres to the same rationale of posting one's emotion in social networking services, such as Facebook. Based on the information of the student's affective state, POLYGLOT delivers motivational messages to the student in support of his/her educational effort. When the student, tries the first test, POLYGLOT receives new information, namely the grade of the student and the time s/he needed to complete the test. Based on this new information, the algorithmic approaches of POLYGLOT may change the affective state of the student and then s/he is presented different motivational messages which adhere to the new affective state. Hence, the student is further assisted and motivated since these messages can indeed support his/her effort. It should be noted that the motivational messages are held in a library and selected every time based on the corresponding affective state.

The presented novel approach of knowledge domain representation and student modeling has been fully implemented in a web-based educational application, which teaches two foreign languages, namely the English and French languages. POLYGLOT is also accompanied with an authoring tool. POLYGLOT's authoring tool allows a non-programmer, usually an instructional designer or technologist, to easily create software with programming features. The programming features are built in but hidden behind buttons and other tools, so the author does not need to know how to program. It provides lots of graphics, interaction, and other tools for educational software needs. The three main components of the authoring system are the content management the type of assessment. The content management allows the user to structure the instructional content and media. The type of assessment refers to the ability to test learning outcomes within the system, usually in the form of tests, discussions, assignments, and other activities which can be evaluated. Finally, it incorporates students' reports and statistics so that the instructor can have a clear understanding of the educational process.

Learning styles are theories that try to separate students by their different and optimum methods of learning. The goal of a learning style model is to find a structure to explain why students have different preferences for learning, and why teaching something one way can be best for one student, while teaching something another way can be best for another student. Individualized instruction is achieved by the use of learning style models because they identify the differentiation and multimodality in the tutoring process. In order to identify the learning styles, it is required by the students to answer a great deal of questions. However, this study initiates the user using a few personal questions about him/her and a machine learning technique to automatic classify them to the appropriate learning style.

Collaborative learning is a situation in which two or more people learn or attempt to learn something together. Unlike individual learning, people engaged in collaborative learning capitalize on one another's resources and skills (asking one another for information, evaluating one another's ideas, monitoring one another's work, etc.). More specifically, collaborative learning is based on the model that knowledge can be created within a population where members actively interact by sharing experiences and take on asymmetry roles. Hence, this study exploits the social networking features, such as digital wall, instant and asynchronous text messaging, in order to provide a collaborative environment and recommend collaborations between students towards promoting mutual learning.

Error diagnosis can identify incorrect learning behaviors, misconceptions the learner may have, and skill sets that need to be developed. It can also be used to determine learners' level of knowledge in between eLearning lessons or modules. Using an error diagnosis mechanism, this study identifies the category of the error that the user made and adapts the learning process by offering personalized advice. Summarizing, POLYGLOT incorporates the following:

- the Stephen Krashen's Theory of Second Language Acquisition
- the Felder-Silverman learning style model
- a supervised machine learning algorithm (k-nearest neighbors algorithm) which takes as input several students' features, including their age, gender, educational level, computer knowledge level number of languages spoken and grade on preliminary test, in order to detect their learning style
- Approximate String matching for diagnosing types of students' errors
- String meaning similarity for diagnosing errors due to language transfer interference
- Techniques for tailored assessment
- the Linear Regression model to automatically detect students' frustration
- the Attribution Theory to deliver appropriate motivational messages to students.
- Tailored assessment to students.

The implemented novel educational system that teaches the English and French languages has been evaluated. In particular, the Kirkpatrick's evaluation model was used and POLYGLOT was evaluated based on its four layers. Particularly, the four levels of Kirkpatrick's evaluation model essentially measure:

- the reaction of student: what they thought and felt about the training
- the learning: the resulting increase in knowledge or capability
- the behavior: extent of behavior and capability improvement and implementation/application
- the results: the effects on the business or environment resulting from the trainee's performance.

POLYGLOT's application was based on close-ended questionnaires and on experimental research. The questionnaire survey was performed in two stages. In particular, two questionnaires were answered immediately after the end of the training program, while the other two questionnaires were answered six months later. The six months waiting time for the follow-up evaluation could have as a result the responses to have affected by students' personal factors. It is known that there is no objective way to deal with it. However, the large amount of students (80) of the experimental group, their answers in the questionnaires of the first stage and the objective experimental research enhance the evaluation results.

The system's evaluation revealed that the automatic detection of the learning style along with the automatic frustration recognition and the delivery of motivational messages contribute, significantly, to the personalization of the learning process to each individual learner. The results of the evaluation demonstrated learning improvements in students and adaptation success to their needs. They revealed that the incorporated error diagnosis mechanism assists the students in the educational process and improves significantly the student's performance. Furthermore, the majority of the learners were very satisfied with the educational program. They obtained a more positive state and behavior towards foreign language learning and distance learning.

10.2 Contribution to Science

Following, the contribution to science in the related scientific fields is presented.

10.2.1 Contribution to Intelligent Tutoring Systems

One important novelty concerning the field of Intelligent Tutoring Systems lies in the fact that social media characteristics are incorporated in the user interface of the learning environment. Social media characteristics, such as posting on a wall, tagging a classmate, instant and asynchronous text messaging, declaring affective state and liking of the exercises, have been included in the Intelligent Tutoring Systems. Furthermore, it uses such features so that the student model is further enriched and the educational process is student-centered. Such features include the following:

- the automatic detection of the learning style based on the Felder-Silverman model,
- the automatic detection of the students' frustration using the Linear Regression model and the respond on this using motivational messages based on the Attribution Theory,
- the recommendation for win-win collaboration and
- the hybrid model for error diagnosis mechanism employing the Approximate String Matching and the String Meaning Similarity algorithms.
- The techniques for offering tailored assessment based on students' needs and preferences.

Finally, one new area in e-learning has been accentuated in this research. When e-learning incorporates social networking characteristics along with intelligence in the instructional process, there is the birth of a new area in e-learning which is called

Social Networking-based Learning (SN E-Learning). SN E-Learning combines a Social Media User Interface with the intelligence of ITSs and as being in its infancy, there is a fertile ground research on this new area.

10.2.2 Contribution to Computer-Supported Collaborative Learning

One important novel module of POLYGLOT regarding the collaboration between students is the win-win collaboration module. The contribution of this module, employing algorithmic techniques, assists students to find the right classmate for collaboration. Win-win collaboration module serves as a recommendation tool which promotes collaboration between students in a way that both of them can benefit from this process. The module supports two different approaches for collaboration. The first one is the win-win collaboration based on the already learnt language concepts. The second approach concerns the types of misconception that the user made. For example, if a student is good at concept A but has poor knowledge on concept B, the system proposes him/her a collaboration with another learner who is complementary to the concepts. Also, under the same rationale, if a student is prone to conduct misconceptions of category A but s/he does not conduct misconception of category B, the system proposes him/her collaboration with a student who conducts misconception of category B but not of category A. As such, based on two significant characteristics, namely the gained knowledge on taught concepts and the type of students' misconceptions, the system recommends collaboration between classmates and they will both learn from each other. To this direction, the system provides advising to learners to collaborate with peers in such a way that both of them can reap the benefits of collaboration and learn while collaborating. The module constitutes an ideal way for collaboration tailored to students' needs.

10.2.3 Contribution to Student Modeling

One of the targets of this research was the automatic detection of the learning style of the student based on the Felder-Silverman model. The target of this research was to offer a more personalized environment to students so that they can learn at their pace, as stipulated by their learning style. The system's evaluation revealed that it contributes significantly to the adaptation of the learning process and to the learning pace of each individual learner. In this way, the presented novel approach helps the learners to save time and effort during the learning process, since the learning style detection is automatic, and to experience a more personalized tutoring process. As such, the learning material is delivered to each individual learner according to his/her learning style, taking into account his/her learning needs and different learning pace.

Furthermore, the hybrid error diagnosis module reveals to the students the type of their misconception in an automatic way and supports them in understanding the gap in the knowledge of the taught concepts. Particularly, the error diagnosis module combined two different algorithmic techniques (approximate string matching and string meaning similarity) into a hybrid approach and supports the user in case of possible confusion with features of the previously-known foreign language. In this way, the system allows each learner to understand the reason of his/her mistake; as such, the student learning can become more effective.

10.2.4 Contribution to Computer-Assisted Language Learning

Computer-assisted language learning systems teach foreign languages to learners, providing adaptivity. Mainly, these systems adapt the learning process dynamically to the student's knowledge level and needs. However, they do not provide automatic inference about the learner's learning style as POLYGLOT does. Consequently, the gain of the presented approach is that it allows each learner to complete the e-training course in a way that the system adapts dynamically to each individual learner's pedagogical needs. Furthermore, POLYGLOT delivers motivational messages to students based on the Attribution Theory in order to support them in their effort and prevent them for quitting the learning. Moreover, POLYGLOT constructs its learning strategy using the Krashen's Theory of Second Language Acquisition which contributes to the field of Computer-Assisted Language Learning in terms of the way of instruction, means of collaboration, time constraints in learning, holding students' records, logical gradation of learning concepts and response on negative affective state (frustration) in the form of motivational messages.

10.2.5 Contribution to Affective Computing

The contribution of this research on Affective computing is the automatic detection of the emotional state of frustration based on students' interaction with the social media user interface and the provision of appropriate response to those emotions in the form of motivational messages. The automatic detection of frustration takes place with the use of the Linear Regression Model which also finds the reason of frustration of the student.

10.3 Future Work

This book presents a social web-based application, incorporating automatic detection of students' learning style using the k-NN machine learning algorithm, two algorithmic approaches for effective error diagnosis, frustration detection based on the linear regression model and motivational messages based on the Attribution Theory. Given that the evaluation results are very encouraging, future work includes the incorporation of other knowledge domains in the system. Furthermore, future plans include the employment of other machine learning techniques, such as Support Vector Machines or C4.5 algorithm, and ensembles of classifiers being based on a variety of classification methodologies and achieving different rate of correctly classified individuals.

The development of a model which adapts the learning content of students based on their affective state and the experimental investigation into whether this model can effectively promote the education process is also an interesting field for further research. The first step is to use the linear regression model presented in this book. The next step is to build a dynamic Bayesian network using associated features for each affective state. One affective state's impact on the other could be modeled as the transitional matrix of affective states. The Markov Chain (MC) can be used for cognitive affinities by the transition matrix and features related to the affective states. In MC, the affective state expressed at the particular point t depends on time $t - 1$ on the affective state. This will enhance student adaptation and personalization. Finally, the hybrid system will be developed using a webcam, microphone, eye tracking system, keyboard with pressure-sensitive keys and equipment to capture students' emotions and further enhance their learning experience.

References

1. Krouska, A., Troussas, C., Virvou, M.: Comparative evaluation of algorithms for sentiment analysis over social networking services. J. Univers. Comput. Sci. **23**(8), 755–768 (2017)
2. Krouska, A., Troussas, C., Virvou, M.: Social annotation tools in digital learning: a literature review. In: 2018 9th International Conference on Information, Intelligence, Systems and Applications (IISA), pp. 1–4, Zakynthos, Greece, 2018
3. Troussas, C., Chrysafiadi, K., Virvou, M.: Machine learning and fuzzy logic techniques for personalized tutoring of foreign languages. In: Penstein Rosé, C. et al. (eds.) Artificial Intelligence in Education. AIED 2018. Lecture Notes in Computer Science, 10948, pp. 358–362. Springer, Cham (2018)
4. Troussas, C., Krouska, A., Virvou, M.: Social interaction through a mobile instant messaging application using geographic location for blended collaborative learning. In: 2017 8th International Conference on Information, Intelligence, Systems & Applications (IISA), pp. 1–5, Larnaca, 2017

The manufacturer's authorised representative in the EU is Springer Nature Customer Service Centre GmbH, Europaplatz 3, 69115 Heidelberg, Germany. If you have any concerns regarding our products, please contact ProductSafety@springernature.com

Printed and bound by CPI Group (UK) Ltd, Croydon, CR0 4YY
05/10/2026
02228069-0001